RICHARD FORTEY spent his _____ the Natural History Museum, specia_____ _____ ning a world expert. He was elected President o_ _____ gical Society of London for its bicentennial year of 2007 and is a Fellow of the Royal Society and of the Royal Society of Literature. He has received the Frink Medal, the Michael Faraday Prize and the Lewis Thomas Prize for science writing, as well as the silver medal of the Zoological Society for science communication. He is the writer of eight previous science and nature books, including two *Sunday Times* bestsellers, all of which are still in print. He has presented many television programmes across the BBC and other channels.

'Truth and courage are what memoirs need and this one has them both in spades ... He never forgets that the small boy, watching his father's effortless casting on the waters of the Itchen, somehow remains permanently present inside the great, famous and lauded scientist. The unforgotten boy: that is what makes this book a revelation'   ADAM NICOLSON

'A wonderful, absolutely beguiling glimpse into the formative life of a great scientist. I learnt a lot and really loved it'
                    RICHARD HOLMES

'A compelling autobiography that shows how an awkward youth became a renowned naturalist ... wonderfully lyrical ... funny and entertaining ... I would also suggest that the real revelation is something other than the way these multiple childhood paths converge ... [but rather] his ability to see and interpret the complexities of the living world, as if from a great height, and then to compress all the technical material into a scientifically accurate form that is also full of poetry and music ...The most compelling insight of the book: the way in which its author has striven to fuse and harmonise, often against career typecasting, professional constraint and simple circumstances, to become the whole person he wished to be ... Both the book and the life it recounts amount to a singular triumph'
                    MARK COCKER, *Guardian*

Also by Richard Fortey

The Hidden Landscape: A Journey into the Geological Past

Life: An Unauthorised Biography. A Natural History of the First Four Billion Years of Life on Earth

Trilobite! Eyewitness to Evolution

Fossils: The Key to the Past

The Earth: An Intimate History

Dry Store Room No. 1: The Secret Life of the Natural History Museum

Survivors: The Animals and Plants that Time Has Left Behind

The Wood for the Trees: The Long View of Nature from a Small Wood

Richard Fortey

A Curious Boy

The Making of a Scientist

WILLIAM COLLINS

William Collins
An imprint of HarperCollins*Publishers*
1 London Bridge Street
London SE1 9GF

WilliamCollinsBooks.com

HarperCollins*Publishers*
1st Floor, Watermarque Building, Ringsend Road
Dublin 4, Ireland

First published in Great Britain in 2021 by William Collins
This William Collins paperback edition published in 2021

1

Copyright © Richard Fortey 2021

Richard Fortey asserts the moral right to be identified
as the author of this work in accordance with
the Copyright, Designs and Patents Act 1988

A catalogue record for this book is available from the British Library

ISBN 978-0-00-832400-1

Typeset in ITC Garamond by
Palimpsest Book Production Ltd, Falkirk, Stirlingshire

Printed and Bound in the UK using 100% Renewable Electricity at CPI Group (UK) Ltd

MIX
Paper from
responsible sources
FSC™ C007454

This book is produced from independently certified FSC™ paper
to ensure responsible forest management.

For more information visit: www.harpercollins.co.uk/green

To Heather and David,
with love and gratitude

Contents

Illustrations

1

The Trout

On the wall of my study a trout in a glass case is facing upstream. Its mouth is slightly agape as it gulps the clear, pure water of the chalk river in which it lived. The fish is not swimming; rather, it is perfectly maintaining its position in a busy current. In life, its fins and tail swished gently to keep it exactly where it wanted to remain, alert for the small prey that nourished it to such generous proportions. Its skin glistens, not in the blatant fashion of the crudely stuffed fishes that decorate some waterside pubs along the River Thames, but sleek and subtle, a slick and a lick of varnish, just enough to suggest the slippery sides of the living animal, but not so much as to point up its artificial afterlife. It is a brown trout, a wild fish, its upper flanks broadly dotted as if by a leisurely crayon, belly silvery white and tessellated with hundreds of fine scales. Nor does it hang all by itself. The preparator has carefully placed model riverweed beneath the fish – not just plonked down, but rather drawn out as if played like hair in a current. The direction of the weed points up the rush of the water, and reinforces the convincing response of the fish to its ambient flow. This is a carefully curated specimen. The front of the glass case is gently bowed outwards and framed by a thin line of gold leaf. At the bottom of the frame – also picked out in gold – 'TROUT 4 lb 13 oz. Caught by F.A. Fortey 13 May 1947 River Gade.' Not long after the Second World War, my father caught this fish from the transparent waters of a stream running through the Chiltern Hills, north-west of London. I was a one-year-old baby at the time.

The trout became an heirloom, taken for granted for half a century. It may have hung in one of my father's two fishing-tackle shops as an encouragement – more likely a challenge – to aspiring anglers. The time came when I inherited it and it became part of the paraphernalia that moves with your family from house to house. On one of these changes of address the removal man was carrying the fish to the truck when he suddenly stopped in his tracks, gawped, and exclaimed: 'Blimey! It's a Cooper!' I had no idea what a 'Cooper' was. 'In your fish world,' said the removal man, gravely, 'a Cooper is like a Stradivarius.' He was something of a connoisseur. Now that it had been identified I discovered a discreet little label at the back of the case: 'Preserved by J. Cooper & Sons, 78, Bath Road, Hounslow'. No commonplace 'stuffing' for a Cooper & Sons fish, what you got instead was 'preservation'. These artists preserved the consummate moments of a fisherman's career. I noticed for the first time the subtleties of taxidermy that lifted this fish from the merely stuffed.

The definitive brown trout curated by J. Cooper & Sons.

That trout must have been an exceptional catch. There is no 'brownie' nearly as substantial swimming in the River Gade in the twenty-first century; water extraction for thirsty Watford and insatiable London has much reduced the flow of such spring-fed rivers originating from the chalk of the Chiltern Hills. The Watford Piscators have not been able to find the equal of my father's fish in their records, so when he caught it I suspect it was a record-breaker. It was worth immortalising by commissioning the Coopers to work their magic, even when money must have been short. In the 1940s, after the global conflict, petrol was expensive and rationed, so a short expedition to the river in motorbike and sidecar from the flat in West London was not to be taken lightly. The Strad of the trout firmament remains a serious item, a treasure.

When I was growing up, trout were important. I spent a lot of time watching them. The River Lambourn runs south-eastwards from the famous small racing town bearing the same name, perched high on the chalk in southern England. The bare downs around the town are lined with post-and-rail fences and open rides, where fast and elegant horses are to be glimpsed from time to time. The humble houses in the town built of tough sarsen stones belie the big money that comes to the owners and trainers of a successful nag. Spring water wells up from the depths to water horse and man alike. The spring-fed River Lambourn is another of those special streams with water as cold and pure as it is pellucid, the natural home of the brown trout. This small river joins the greater River Kennet at Newbury, and a few miles north-west of that town I first remember watching fish. In the little village of Boxford I hung on tiptoe over the flint-and-brick bridge looking at the Lambourn trout below. The stream flowed fast here, and just as in the Cooper & Sons 'preservation' the trout often held a position facing upstream. The stream bed was verdant with water weed: crowfoot with dark, thin leaves making bunches that waved like windblown locks of hair in the current, starwort growing in cheerful green cushions on muddier patches in still water near the bank, and stiffer sedges that bent erratically under the bidding of the water's flux. The spots on the trout disguised the fish against their weedy background. They could sometimes almost disappear, and then reveal themselves suddenly with a twitch of the tail, or by a sudden realignment in the current. If disturbed by a tossed stone they might shoot off to lurk under a bank or negotiate a deeper pool. From time to time a grayling would appear as a silvery flash, blatantly displaying its prominent dorsal fin. The river and its fishes were an endless source of variety while essentially forever remaining the same. The timeless emotion that consumed me was similar to contemplating a real fire in a real hearth: flames lick and mutate, and there's nothing to distract you except endless distraction.

I spent quite a chunk of my childhood half wild in the countryside. Although we lived in the West London sub-urb of Ealing, most summer weekends the family escaped to Berkshire and to the fly fishing. Even now I wonder whether I am more urban or more rural. By the time I was eight or nine my father had joined the Piscatorial Society, which allowed him to fish several chalk streams besides the River Lambourn: a stretch of the River Kennet not far from Newbury, and a piece of the River Itchen just outside the ancient city of Winchester. I had a chance to follow the dedicated fisherman along the banks of these famous rivers, honing my skills at spotting fishes lurking under the waving fronds of weed, or observing one of them occasionally rise to sip a fly from the gliding surface of the water, as delicately as my maiden aunts tasted Sunday afternoon tea; hardly a sound, just a small suck of the lips. The skill of the dry-fly fisherman is to place the right kind of artificial fly (with its hidden hook) just where the trout will be tempted to take it. My father was consummately skilful at this. He could place a 'fly' on a penny. The art was to keep the 'fly' aloft at the tip of the thinnest part of the line, by elegant aerial swishes back and forth, until just the right length of line had been conjured from the reel to drop the 'fly' on to the point in the stream where it would pass over the lurking fish. If the trout were tempted to rise to the bait above it the angler must strike at precisely the right moment to snare the hook in the fish's mouth. A moment too soon and the fish will scoot away. Larger fish had survived the attentions of many eager anglers to become more cautious – so the bigger the prize the harder the catch. Record fish have been scarred by experience. The luck of the Coopers' 4 pound 13-ouncer from the River Gade finally ran out when my father came along.

The imitation trout flies used on chalk streams are not the gaudy lures beloved of salmon fishermen that bear little resemblance to real, living insect species. Instead, the standard 'flies' of the chalk-stream anglers bear modest, understated names like Pale Watery or Olive Dun. They are true approximations to real species, and they are comparatively tiny – little tan-coloured scraps of things. They are not even flies in the scientific, entomological sense – they belong to a different insect order (Ephemeroptera*) from houseflies and their prolific ilk (Diptera). Savvy old brown trout are not easily fooled by show alone; they need the real thing. Chalk-stream trout anglers regard their piscatorial skills as the quintessence of the art. To be a great trout fisherman on these

---

* Ephemeroptera are unique among insects in having an extra growth stage. After they hatch out from the subaqueous nymph stage, a brief early 'dun' winged stage moults once again to give rise to the reproductive 'spinner' stage.

challenging rivers is to understand the psychology of the fishes, the entomology of their favourite snacks, the physics of flow, and to match that understanding with flawless skill in casting and striking. Eventually landing the fish is the least of it, although wily and experienced fishes have been known to escape even after they have taken a hook by scuffling through the weeds. A certain snobbery on the part of skilled trout aficionados is perhaps only to be expected, and fishing rights along elite stretches of river do indeed come with a huge premium. The River Test in Hampshire lies at the apex of this hierarchy and the Houghton Club near Stockbridge lies at the apex of that apex. This is arguably the most exclusive club in the world, where a fortune alone will not buy you entry without a deft command of the subtleties of fly fishing. My father had the skill, but never had the money.

There were a few days every year when all the decorum of chalk-stream fishing went to the wall. When the mayfly hatched trout went berserk. This 'fly' was substantial – half a dozen Olive Duns rolled into one fat body. After a year of fattening up as larvae on the stream bed the mayfly simultaneously attained their adult form to fly and mate in the warm air of early May, an adult life of just a day. These ephemeral insects had been doing just this since the time of the dinosaurs. During a good year the air was thick with mayflies, dancing up and down. They landed on outstretched arms like transparent-winged butterflies. This small boy examined them minutely, observed how their wings folded back as they landed, noticed that there were two long whip-like hairs at the end of the body, and tossed them into the air again to see them dancing. When they landed on the water even the shyest, most reticent trout could not resist the succulent bodies of breeding mayflies. Now the gulps of the fish were readily audible, and occasionally a whole shining body churned and twisted in pursuit. Fly fishermen look forward to mayflies as small children long for Christmas.

If William Wordsworth was right that 'the child is father of the man' then I was first made on a riverbank. Did my early life, linked so intimately to trout, somehow dictate the person I became? That small boy in grey flannel shorts doggedly tracking a virtuoso father along a chalk stream eventually became a scientist and writer. The roots of my own life were nourished on summer days in the Berkshire countryside just pottering about and wading through marshes. My recall of that young boy is often vague. However, I do *precisely* recall early sensations and observations: the slightly tacky feel of the mayfly limbs stuck to my arms, the dank smell of waterweed, the frenetic scuttling of freshwater shrimps. As far as I know, no other boy from Ealing had such a curious connection to chalk streams. They didn't have freshly fried brown trout for breakfast. This was almost the only cooking my father ever did: brown trout fried rapidly till the skin went golden, but all the luscious taste kept inside; nothing could have been more scrumptious.

I did not savour that taste again until thirty years later when I persuaded a young man to part with his catch of sea trout in western Newfoundland. Then that special deliciousness immediately caught the past with poignant accuracy. I remembered how I used to fillet the trout myself, removing head, guts and swim bladder in a single cut, and how a slippery fish once shot from my hands as if it were still alive. If I am to explore how that small boy became a scientist I have to follow prompts like taste or smell, revive half-forgotten memories, stir some murky depths. There are no diaries for me to refer to, and the exact chronology of many events has faded over time. A few important moments stand out as transformative markers, but I cannot lay claim to a sense of inevitable destiny. I did not dream of cracking the secrets of prime numbers when I was ten, nor did I anticipate the discovery of a new fundamental particle when I was thirteen. Serial enthusiasms were more the ticket, and I still believe in the inexhaustible interest of the natural world. There were times when happenstance determined my future direction, and I wonder what would have changed with a different spin of the coin. What I do have as tangible evidence is that 'preserved' fish from Cooper & Sons, a physical prompt to conduct me back to the past. There are other portals that open on to my younger self – a scuffed book or an ammonite – that trace how an instinctive naturalist developed into a professional palaeontologist; but that small boy wandering along the riverbank is still in there somewhere, always waiting for something new to turn up, as alert as a kingfisher with a minnow in its sights.

The temptation is to portray those riverine days as somehow bathed in sunlight and serendipitous discovery, but there were difficult moments. Fishing beats leased to the Piscatorial Society were policed by gamekeepers with relentless vigilance; they took no prisoners. By the River Kennet damp pathways led between coarse tussocks of sedge and treacherous ditches full of reeds that were far taller than the boy wandering from one distributary of the river to find another. Suddenly, a gamekeeper's 'larder' loomed up without warning – a crude wooden contraption on which the guardian of the fishing had hammered the bodies of predators he deemed a menace. Weasels and stoats endured a ham-fisted crucifixion, magpies drooped like wounded umbrellas, crows hung limp, all lustre faded from their feathers. But the larder was alive: the corpses heaved with maggots, while big black beetles climbed the scaffolding of the corpses in search of some unclaimed patch of putrescence. The stench made me reel. In a horror movie the scene would have been announced by a sharp discord from full orchestra. But I was curiously transfixed by this tableau of death. I suppose the theory was that such a grisly display would deter predators. It certainly deterred one small boy. I don't know how long I stood there, but the spell was broken when a wing of one of the birds just fell off, and several creatures scuttled on spindly legs from the fallen limb into the surrounding grass.

Kennet evenings were a different matter. Outside the mayfly orgy the favoured time for a trout fisherman was, and remains, the 'evening rise'. It seems that trout like to taste a fly or two as the sun dips below the horizon. A strange stillness embraced the river at this time. Moths emerged from their daytime hiding places and fluttered all white and silent over the water, which assumed a profound and almost greasy blackness as the light left it alone. The swish of the rod accompanied the faintest gurglings of the restless stream. If there could be such a thing as perfect peace, then this was it. The rise was most satisfactory when the weather was clear and warm, and there were occasions when the prospect of an interesting evening would entail a hectic drive from London to catch the fading light. My father habitually drove too fast. We would, however, always have to stop the car once, and sometimes twice, to wipe dead insects from the car windscreen. This simple plate of glass was a way of randomly sampling what was flying as dusk approached – and there was a lot of it. The most conspicuous insects were obviously succulent moths, but there was also a fuzz of lesser beasts: mosquitoes or gnats, perhaps. Thousands of them.

That is what has changed. When I drove down some of the same roads recently, in the dying light, I arrived at my destination with perhaps a dozen insects on my windscreen, and only one moth. Maybe the aerodynamics of windscreens have changed during my lifetime to kill fewer insects, but I am certain that their flutterings do not catch the headlights any more, regardless of whether they finish up as dead bodies. If a symbol were needed of the decline of the country habitat since the 1950s then a clear windscreen would be hard to beat. The recent reduction in the number of insectivores of many kinds hardly comes as a surprise. Blame is easy to dish around, but the hegemony of monocultures on ranch-style farms and the liberal use of insecticides is going to be on the list of prime suspects. Sometimes, a long memory is a recipe for gloom.

What of the River Lambourn? In 2015 I went back to Boxford, a mile or two north-west of a now expanded Newbury, to revisit the chalk stream that had nourished my first awareness of the richness of life. The old bridge was still there, with the clear, rushing stream beneath it. A grey wagtail welcomed me back, bobbing and flitting like a coquette. The village had changed completely. When my parents bought a simple, thatched cottage in Boxford almost sixty years ago there was no mains drainage. Primrose Cottage was pretty enough, but we did our business in a closet, sterilised by something called Elsan, and then occasionally buried the digested produce in the vegetable garden. I recall the lush gratitude of the cabbages. Nowadays, Boxford is as smart as Chelsea. It looks positively *groomed*. The hedges that surround the old cottages are shaved as close as a glamour model's armpits. Somehow, the village no longer feels like part of the countryside, more a lifestyle accessory. However, while the clean river still flows vigorously, all is not lost.

I had joined a team from the Environment Agency that has been monitoring the health of the River Lambourn. Boxford Meadows is now a Site of Special Scientific Interest. One of my early sources of biological wonder now has the imprimatur of an official classification. In wellington boots I followed the team as they counted the fish that were retrieved from a stretch of water they had selected for sampling. Crack willows still leaned precariously over the rippling stream, and the water crowfoot was sporting its white, buttercup-like flowers in profusion, each bloom held aloft from the water to be pollinated in the air just like any other flower. The good news was that there were still wild brown trout in the Lambourn. The biologists used 'electric fishing' to take a census of the fishy life. Place electrodes into the water from a small punt and native species cannot resist being attracted to the positive pole. The lure of electricity turns them into zombies drifting towards the electrode – where they soon become statistics, before being returned to the water unharmed. Along with the trout I saw the brook lamprey (the lampern), a fish so primitive it does not even have a proper jaw. It looked superficially like a small eel, until I noticed the strange lines of gill openings behind the head. Like the mayfly, it is a fugitive from deep time. Some species failed to turn up, like the 'miller's thumb' (*Cottus gobio*), a small, fat-headed, spiky fish I caught long ago using a small net poked between the flint pebbles that covered the bed of the river. Then an unwelcome stranger barged into the inventory: the American signalling crayfish, *Pacifastacus leni-usculus*, as big as the palm of my hand. When I was a child, a smaller, native, white-clawed crayfish (*Austropotamobius pallipes*) was there instead. I remember picking one up by the carapace just behind its claws, which opened like miniature nutcrackers when the creature left the water. Our own modest crustacean has since become virtually extinct, bullied by its distant cousin from the other side of the Atlantic Ocean. The signaller was introduced to Britain in the 1970s as a 'cash crop' and has since relent-lessly advanced into all streams with pure running water, where it has exterminated its native ecological rival. It is voracious – even cannibalistic – and I cannot help wondering whether the absence of the 'miller's thumb' might also have something to do with this interloper.

Nor was every trout a 'brownie'. A few rainbow trout among the sporting fish identified themselves by red stripes running along their flanks and dense spots above. The rainbow is the farmed trout familiar from every super-market. They originated from cool streams in North America that flow into the Pacific Ocean, and they, too, have become naturalised in Britain. Many trout-fishing clubs stock their rivers with rainbow trout grown in stew-ponds to give their members an easier catch: they grow faster than the native species, and raised fish are less wily. My father rather despised them. He believed only brown trout truly tested the artistry of the fly fisherman, and the wilier the better. We rarely ate rainbow trout. The soft pink flesh of the wild brown trout, nurtured on freshwater shrimp and flies and snails, was considered a far superior food. It was many years before I could bring myself to purchase a pack from a supermarket, by which time my memories of the wild fish from the chalk streams had begun to fade; but I still knew something was missing.

My day by the River Lambourn left me feeling curiously empty. I had first been encouraged by the integrity and enthusiasm of the young scientists monitoring the stream, which still retained its transparent fascination for me a lifetime later. I could hardly expect it to be unchanged, crayfish notwithstanding. Then, too, so many wild flowers had gone: my sister and I used to gather marsh orchids and ragged robin from the water meadows, when these plants were so profuse that the boggy fields were coloured

red with them. I could see that the young biologists did not really believe me when I told them about it. 'Rose-tinted memory,' I imagine them thinking, although it was as true as my recollection of the crayfish. My sister was tiny and her bunch of flowers was enormous. Nowadays, we would be horrified by the thought of picking such recherché glories, but it was not the thoughtless picking that prevented their survival – these plants had been thriving on the water meadows for generations. Changes in drainage and the addition of artificial fertilisers to increase yields are more likely culprits. How, I wondered, can you properly convey such loss. When I see those rich leas in my mind's eye they provide more vivid and solid memories than those of my own father.

Ambivalent feelings did not stop me from further explo-ration of the Lambourn Valley: I wanted to know the worst. I followed the minor road that runs alongside the stream north-westwards towards Lambourn town, through the villages of Welford and Great Shefford. Now the whole valley had been tastefully prettified. The original cottages were built of cob, or brick and flint, both humble materials that did not have to travel far. The chalk country offered no natural tiles for roofing, so thatch, much like that on our Primrose Cottage, was still a common sight. The river is dwindling. It is easy to follow its course, as lines of willows track it faithfully at the bottom of the valley, their roots nakedly plunging into pools and rills. But somewhere north-west of Great Shefford the streambed seems to have run dry. Maybe it flowed in the winter, and disappeared in the warmer months: there were still trails of flints that marked its course. In Lambourn itself I saw the channel where it was *supposed* to run through the town, but the stream seemed to have deserted its name-giver. The water table must have fallen to the point where the upper part of the famous chalk stream had run dry. It reappeared downstream where the water table reached the surface, and by the time it reached Boxford the Lambourn was approaching its old self. What was happening now in Berkshire had probably happened some years ago in Hertfordshire to the River Gade, from which a great trout had been snatched a few months after my first birthday. Extraction of the chalk aquifer for too many thirsty throats caused diminution of both rivers. People like to drink chalk water because it has been cleaned and purified by its passage through the white limestone rock: it tastes good. Loss of animals and plants is the cost of tap water *ad libitum*.

The Lambourn taught my sister and me to swim. The Piscatorial Society 'beat' on the river extended south of Boxford to its lower reaches at the village of Bagnor, very close to Newbury. An old mill house interrupted the smooth passage of the stream and below that disused building a mill pool was fed by a cascade that fed through a weir. The river spilled out on the other side into shallows with many a 'miller's thumb'. The mill race was fast, but fed into a great deep eddy, so that a floating object would be carried around it in a circle. Along one side of the pool was a submerged ledge. Spring-fed water straight from the chalk was allegedly the coldest in southern England, and every fresh immersion in it provoked a squeal. On perfect summer days it was the equally perfect place for a picnic on the bank, while my father was off with his rod. Under our mother's eye, we children started with rubber rings around our waists, and jumped from the ledge into the mill race to be carried like flotsam around the eddy to repeat the experience all over again. At some point the rubber rings were shed – and we were swimmers. My sister Kath was two years younger than I was, but she learned to swim first; I can still feel the touch on my feet of the special moss that grew on the mill race and I can still smell the strange exhilarating odour of pure rushing water. Once, after swimming, mother and children went to a scruffy pub on the far side of Bagnor village green. My mother was extravagantly flattered there by an aged charmer, who insisted on buying her a drink – and an expensive one at that. At that time she was at the peak of her womanhood, tall, fit, and often wearing her 'sundress', a light flowery confection that exposed her tanned shoulders and flattered her figure. Before this, I had never realised that she was a sexual being, which is probably why this incident has lodged so obstinately amid the general fuzz of my growing up. Afterwards, she giggled and referred to her admirer as an old fool.

The old mill is now the smart Watermill Theatre. It offers a judicious mix of classical plays and popular shows, and an evening out (with supper included) for the people of Newbury. The mill race is still there, but it is hard to believe now that the wildest of wild swimming happened in its frigid waters. When I visited recently I thought I caught a glimpse of a native trout in the shadowy depths. The marsh orchids have declined, and the water meadows are now little more than civilised places for dog walking. Nobody seems to remember that this was once an untamed place, alive with insects and warblers.

*   *   *

Fishing was not just for the weekends. It was my father's business. He had two fishing-tackle shops in London: Tooke's at 614 Fulham Road, and Harding's, in Willesden High Street. The names of the previous owners were retained because anglers are a conservative lot, and the shops themselves were survivors from before the war with nostalgia as a selling point. Neither shop was in a smart part of London. The Willesden shop seemed to belong to a London that had vanished from elsewhere in that city, a dim cavern of a place that Dickens might have peopled with a minor character bearing the name Mr Crepuscule. The Fulham premises was much smarter, with a glass shop window displaying desirable fishing rods, Hardy's reels, and keep nets. Inside, glass-topped display cases showed off a selection of the angler's accoutrements, while against the wall fishing rods were lined up vertically, ready to be demonstrated to customers wishing to find the magic recipe for the big fish. There was a sign saying 'live bait'. Fulham is now a middle-class area, largely populated by City workers, but in the 1950s it was working class; and parts of Willesden were even considered rather rough. The rods on sale were predominantly for coarse fishing. The arcane refinements of the trout rod were rarely discussed, because the items on display were there to hook roach, carp, bream, chub and the mighty pike. The customers did not parade along the banks of the River Test or the Itchen. They were Thames fishermen, or canal fishermen, or went on expeditions to the Lincolnshire 'drains' to hunt out giants. Angling was the great sport of the working man. The two shops had local managers, known to us children as Uncle Arthur (Willesden) and Uncle Eddy (Fulham), who wore brown coats, and addressed the customers as 'squire'.

The rear room of the Fulham shop was a different world. The walls were lined with glass-fronted fish tanks, and freshwater tropical fish were on sale. There was a constant bubbling sussuration as the tanks were kept oxygenated. This part of the shop was mostly dark, so that the lights that illuminated the tanks pointed up their exotic occupants. Green waterweed decorated bigger tanks, and they became small stage sets, through which the fish played and darted. I learned the cast of characters. Zebra fish were small, black-and-white striped, and active – but I noticed that the stripes ran the length of the fish rather than zebra-like down the flanks. The name felt slightly wrong. Guppies bred in some tanks, the little female fish so dowdy compared with the male, with his tail like a gaudy streamer. The female gave birth to live young, and the tiny fry hid among the weedy fronds. Diminutive and slender neon tetras had the most extraordinary luminous blue stripe on the side of their bodies, as if they were lit from within. Angel fish had very deep bodies and moved slowly, while black mollies were busier, regular-looking small fish but so dark they could look like moving, fish-shaped holes in space. Swordtails carried red 'swords' on the lower part of their tailfins. Tiger barbs had a few black slashes down their bodies – hardly a piscine tiger but the stripes did at least run in the right direction. Lilliputian, dark catfish cleaned the gravel incessantly. Gouramis were larger, deep-bodied fishes aimed at the connoisseur, decked in stripes of all colours, or spotted in interesting ways. Some of them were expensive. The king of them all was the Japanese fighting fish, with fins as spectacular as a flamenco dancer's twirling skirts. These fishes had to lead solitary lives because they were so territorial; if another male was introduced battle commenced. I spent hours closeted with these tropical fishes, and it may have been in their company that I acquired the habit of taxonomy, the urge to name names founded upon close observation. I got some idea of the diversity of life, of the way nature could play around with bodies and colours, and how there seemed to be no end to natural invention. I may, or may not, have known about evolution; I cannot remember whether or not the Cooper trout was hanging in that same back room. But I remember the angel fish.

My mother, who was very aware of social distinctions, urged me to describe my father's profession as that of 'aquarist' when talking to the mothers of my friends in Ealing. She thought it sounded far more acceptable than 'fishing-tackle dealer'. 'Trader in maggots' would have been truly humiliating, but that is what happened on Saturday when the anglers came into the shop for their bait. I lent a hand at busy times in Tooke's, serving quantities of writhing maggots to paying customers. These grubs were known in the trade as 'gentles' and were served in half-pint measures – I had to dip the measure into a bin containing thousands of fly larvae and dish them out into boxes with perforated lids that let the creatures breathe. They came plain or coloured – the angler could pay a little more for dyed pink or yellow gentles. An adjacent bin contained worms – bright red ones – favoured as succulent bait by many customers. Other anglers liked to 'dope' a popular patch of river with broadcast hemp seed to attract fish towards their baited hooks – this seed was stored in yet another bin. I have no idea whether the hemp seeds would have grown into cannabis plants. The fact remains that my first commercial experience was as a maggot dealer, with hemp on the side.

My father once took me to pick up his supplies of maggots and worms. He drove our Austin to a decrepit farm on the western outskirts of London, where the countryside was slowly giving up the struggle against being built over. Scrub was commonly the only crop. The supplier to the trade was known as Wormy. I suspect he had no other name. Wearing an ancient cap, and grinning through snaggled teeth, he took us to a kind of industrial gamekeeper's larder, a low shed with lines of condemned meat corpses dedicated to raising the gentles, which were collected into trays as they fell off. The buzzing of flies was overwhelming. Regular red worms were raised in compost heaps fed with cow manure. They could be grabbed by the shovelful. Wormy loaded the next week's supplies into square tins with tight-fitting lids. It was important to the anglers that the gentles should be vigorous wigglers of a similar maturity. If they were kept in the shop too long, they slowed down and pupated, on their way to becoming yet more flies. Money changed hands ('cash terms only'). I don't remember saying anything on the way home.

Gathering fresh food for the tropical fish was a much more pleasant routine. Walpole Park in Ealing, the 'queen of the suburbs', was close to our London home. Some years later I would go to the school right next door to the rose garden, which was part of the park surrounding Pitzhanger Manor (a mansion built by Sir John Soane, the notable architect, in 1800–4). The large manor park passed into council ownership, and remains so today. On one side of it a landscaped pond was home to waterfowl and entertained young families. It also supported uncountable numbers of tiny water fleas (*Daphnia*). These minute crustaceans often flourish in fresh water that is low in oxygen, and in Walpole Park there were so many that they coloured the water pink. The tiny creatures manufacture their own haemoglobin, the same component as in our own blood, helping them get the oxygen they require to thrive in an unfavourable habitat. In the water they were no more than masses of tiny, twitching specks. My father took out a very fine but capacious net on the end of a long pole, the handle of which was made of several lengths screwed together. He swished the net back and forth, harvesting thousands of the water fleas. They made a kind of sludge at the bottom of the net, which he tipped into a large jar filled with water, brought for the purpose. Then back to the pond again for another haul until the jar housed a dense, living pink broth. From Walpole Park we went down to the River Thames near Brentford, just a couple of miles away. This was before London's river had been cleaned up, and most aqueous life struggled to survive in its soupy waters. At low tide a strange and unpleasant smell assailed the nostrils, while banks of glistening dark mud were exposed, shining sinisterly in the afternoon light. Pulling on wellington boots my father marched along a greasy rivulet that crossed the mud towards the river and proceeded to scoop up handfuls of mud that he immediately dunked into another jar. As the water cleared he showed me that the harvested mass actually consisted of thousands of tiny, thread-like worms: *Tubifex*. They writhed like tiny versions of the myriad snakes that crowned the Gorgon, and they were red, coloured the same as the water fleas, and for the same reason. They had developed red blood to cope with the foul conditions of the Thames mud. Where other living things found life unbearable, for *Tubifex* it was an opportunity to prosper. I would take that lesson on life's exuberant opportunism into the rest of my scientific life.

When *Daphnia* and *Tubifex* were fed into the tanks back at the Fulham shop the tropical fish went into a feeding frenzy. They instinctively knew that this was the food of the gods, although some of them had probably never encountered such living delicacies before, after being raised on dried fish food. *Tubifex* worms thrashed ineffectually as they were gulped down. The fishes darted through the weeds to find every last water flea; up close, I could see the beating limbs of the tiny crustacean that jiggled its little carapace before it was sucked into the open mouth of some small tetra. Aquarists knew how to give the favourites in their collections a proper treat, and they bought small quantities of live food from my father to take home: it was pure profit. However, it was profit based upon knowledge.

Harding's of Willesden was a dark cavern of a place. I think my parents must have purchased it for a bargain price, not only because of its general decrepitude, but also because upstairs there was a sitting tenant, Mrs Oakins, whose rent had been fixed in about 1922 at half a crown a week (about £7 in today's money). She seemed to me to be very old, and ran a spiritualist circle from her parlour. It would not be long before my parents had vacant possession (or so they said). Mrs Oakins disappointed them by living for ever: indeed she *was* very old. Her hundredth birthday came and went. For somebody who was supposed to have such a conviction in the afterlife she showed a remarkable reluctance to go there. The back of the shop had a small and dingy room that was another of my father's redoubts. While Uncle Arthur tended the shop during the close season, when the trout are protected to allow them to breed (October to March on chalk streams in 2019), my father would disappear into the gloom of the back room and craft fly-fishing rods. Modern fishermen use rods made of sophisticated carbon fibre, but split-cane was high technology at the time of Mrs Oakins. These rods were bespoke: they were made for special customers, even the kind who might have fishing rights on the River Test.

I still have one of them. The fishing rod comes in a brown cloth bag in three pieces for reassembly at the riverbank; above the cork handle is the legend 'Fortey Special'. It is put together from cane slivers glued together to make a hexagonal rod about ten feet long and tapering to something hardly wider than a matchstick at its apex. The glue was a most horrible smelling product made from animal bones that had to be kept warm on a danger-ous-looking gas ring: naked flame amidst papers and paraphernalia. The glue pot had congealed dribbles of coffee-coloured gloop around its rim. No doubt modern safety legislation would have made the whole process illegal, but legal or not there was something timeless about the intense work needed to complete the piece. Once the basic rod had been assembled, clamps were removed, and ferrules to guide the line were fixed in place at intervals along its length, bound by whipped cotton. Hollow brass joints were inserted to allow the three pieces to conjoin. The whole rod was varnished to a high finish. Some of the rods were for 'wet fly' fishing, where the idea is to catch the fish by the ruse of imitating the fly nymphs rather than adults. Ephemerid flies like mayflies had larvae that matured on the streambed, but finally came to the surface as adults briefly to fly free and propagate; the nymph was a halfway stage ready to come to the air but not quite liberated from the water. The imitation nymph had to be just under the surface (hence wet fly) rather than float above the target of its attention. This kind of fishing suited small streams and hard-to-reach pools. The customer list has long gone, but there is a note from a very smart gentlemen's club on Pall Mall thanking Frank Fortey for his 'Special' and congratulating the maker on its suitability for a particular trout stream. I once saw my father pocketing a number of five-pound notes for one of his rods – those banknotes were rare, white things approaching the size of a small handkerchief. This was real money, almost a fortune.

Most of the money from the sale of regular fishing tackle was made in the two shops at the end of the close season for coarse fish, which usually fell in the middle of June. After three months of deprivation much of the (overwhelmingly male) angling population of London could hardly wait to get hold of rod and line. The latest models of fishing reels, or shining spinners, or even cheeky floats disappeared in short order. Wormy must have been working overtime to satisfy the demand. In the Fulham Road there were queues. For our family it was all hands on deck and I was kept busy with the maggot and hemp measurers, while the 'ching!' of an old-fashioned cash register hardly let up all day. I learned that money really did have a smell. The only problem was that fishermen are naturally garrulous, and once they got the attention of Uncle Eddy or my father they would launch into tales of past triumphs – even of the ones that got away – oblivious to the desperation of those further back in line. Attempts to move the raconteur on his way were accomplished by deployment of cunning clichés. 'Say what you like, squire, it's a funny old game,' my father would intone, drawing a conversation to a swift conclusion, 'Next!' My mother even made an appearance behind the glass display cases, though she never called any customer 'squire'. The back room with the tropical fish was usually unpopulated, and I might slip in there for a few moments to watch the Japanese fighting fish on the lookout for an invisible rival.

When trout were off limits my father went coarse fishing. He briefly held the British record for the biggest dace. During the 1940s he was the champion angler of the Watford Piscators, and the local newspaper showed him struggling to carry home a plethora of silver cups from their annual dinner. By the end of the Second World War many of the old canals were falling out of industrial use, but had become decent places for fishermen. The old Kennet and Avon Canal still held water for much of its length, and near the village of Kintbury my father attempted to teach me the technicalities of angling. There were roach to catch, and perch that could even be taken to the table, the occasional pike deeply lurking. The path along the canal was built for horses to pull longboats carrying loads of coal, but now the very occasional working boat that passed was diesel-powered. Fishermen had scooped out little nests at intervals among the lank plants along the bank, usually at some distance from the locks. Passing walkers would enquire: 'Any luck?' On this canal bank I learned that I would never become an angler. Like the literary critic who never writes a book, I am fluent in the language of fishing, and love it with a passion, but I am not, and will never be, a successful practitioner. I have devised scenarios to justify this failure. Does the son of a violin virtuoso take up the same instrument? Rarely. The truth is that I am cack-handed and easily discouraged. My attempts at casting resulted in tangled lines; I could effortlessly hook a waterside thistle but miss the large stretch of canal before me. I mysteriously landed a bacon sandwich hitched from the lunch box. Even when gentles and line made it into the water my own float was motionless while a few yards away my father's was bobbing furiously every few minutes. The fish were filed in a large keep net, to be released later, but I fumbled with the little gadget that removed the hook from the fish's mouth after it had been landed. Fishing really was the most complicated procedure, no question, and this was not even the fine art of landing trout. If I couldn't scramble up the foothills, how could I tackle the mountain beyond? Nothing was said. That was usual with my father; but he must have been gravely disappointed. He would have started his angling life on the River Severn in Worcestershire when he was a similar age to me, and I know he must have been effortlessly good at it at once, as he excelled in all sports. His sporting skills had taken him from a small village near Worcester to an Oxford college. He had played serious hockey. He was boxing champion at Royal Grammar School Worcester in 1923, and twice school champion at fives. During the Second World War he had taken up golf, and acquired a fantastic handicap in no time at all. If he had played marbles it would have been at Olympic level. My mother was a good tennis player, and as strong a swimmer. Some quirk of genetics determined that these natural abilities passed me by. I even struggled to take a hook out of the mouth of a small perch.

If I failed to be the angler my father might have wished for then I was ready enough to absorb his knowledge of the countryside. He was a naturalist of the old school, with a broad catalogue of birds and wild flowers; he knew the signatures of the seasons. He used old English titles: the common lapwing was a 'peewit' to him, and the waterside willow herb was 'codlins and cream'. I learned the mysterious poetry of proper names. I discovered that to identify and name plants or animals was, in a curious sense, to own them. Names were sequestered from the endless complexity of the living world to bring some sort of order. A name was more than a dry label that might be stuck to a specimen in a museum. It was as personal as an entry in a directory. Without names, the Lambourn swamp would have remained a vague muddle of collaborating plants and animals, a kind of generalised swirl of existence. With identifications added, it could make an ecosystem – even though that term was hardly current when I made my first guesses. I instinctively knew that naming was the first part of understanding. According to the Book of Genesis Adam named all the animals before Eve was created: evidently, the ancient scribes appreciated that taxonomy provides the key to grasping the world. Without such a foundation, humans wander blindly in an unstructured wilderness. A stream bank happened to be my baseline for cataloguing nature. Had I been born into a desert society I might have been fascinated by arrangements of thorns or anatomised the footprints of lizards on the sand. When I visited a South American rainforest with a local guide I was impressed by the long litany of species in his native tongue – every kind of bird and tree with its own identity. Species are the general units for getting a grip on the natural world. Maybe

I also came to understand from early on that this primal task of recognition differs from *classification*. There are many ways to organise species: according to their edibility, or use in medicine; their diet, colour, taste or smell. The scientist prefers to structure his or her classifications to reflect the deep processes of evolution. Before such rationalisation is possible there must always be a process of making close observations, to satisfy curiosity, to answer the question: what exactly is this organism? I have continued to chase this simple question through much of my life as a naturalist and scientist. It must have first flashed into my mind somewhere near a brisk trout stream.

*   *   *

The Piscatorial Society owned fishing rights on the River Itchen, a fine stretch of chalk-fed water in Hampshire; it was the furthest of their beats from London, and my father did not get there as often as he did to the Kennet or the Lambourn. The stream's clear waters ran directly into the ancient town of Winchester. It was strange to turn off an old city street immediately into a quiet field, where nothing could be heard except the licking sound of fast-flowing water. The city was founded by the Romans in about AD 70, when it was known as Venta Belgarum, capital of the local 'Celtic' tribe, the Belgares. When I studied Latin at school I was compelled to read the text of Caesar's *Gallic Wars* (Book 3), which was stuffed with lists of the tribes the mighty Julius subdued one after the other: the Nantuates and the Sotiates float upwards from some deep recess of my memory, at the head of a long catalogue. The Celtic Belgares must have been yet another, later feather in the Roman cap. This outpost of the Roman Empire provided my first taste of archaeology. Close to where the anglers parked were piles of stones and debris brought up from a recent dredging of the riverbed. Two giant oak piers lay among the miscellaneous heaps. A local archaeologist who talked to my father was convinced that these were the remains of the Roman gates. In the second century the Romans had fortified Venta Belgarum with walls, and robust wooden gates halted visitors and traders before they could enter the prosperous city. These two giant, shaped trunks, now bleached with age, may have been the uprights that supported the gates to the old town. I do not know whether these speculations ever got the blessing of a scholarly assessment. What I do know is that from the debris I retrieved small pieces of ochre-coloured ceramics that really were from the era of the Romans. Humble tiles, maybe, but direct evidence of the ancient ruling power. There were fragments of glass grown misty with time, and how desperately I longed to find a coin! By the end of the day I had a small collection – the first collection of many in my life – and I had also gained some notion of antiquity. Time, I realised, had depth; there were vanished worlds to explore.

In the fading light of the evening rise, a natural soundscape accompanied the fisherman's slow tread along the bank of the River Itchen. The repetitive high bleat of the coot was almost startling as the black bird fussed around its nest concealed among the yellow flags. The hoot of an owl quartering the water meadow was not remotely spooky; it was more like a greeting. The most frequent sounds were gentle splashes into the stream from the bankside. Water voles were abundant along the Itchen in my childhood. These black rodents diligently swam back and forth across the river with their mouths stuffed with damp vegetation, rather like small aqueous retrievers, before disappearing into tunnels in the banks that led to their nests. My father even cursed them for undermining his fisherman's beat; the bank sometimes wobbled alarmingly if the voles had been too thorough. The water vole was Ratty from *The Wind in the Willows*: no rat at all, of course, with its short face and furry ears. Rat and vole really only share a common gift for incessant activity: the vole vigorously chomping its vegetarian diet, and swimming largely submerged as busy as can be, and as harmless as it is blameless; the brown rat up to no good with anything edible, and spreading disease at a brisk trot. There was plenty of evidence for Mole, too, along the waterside. Molehills dotted the grassier patches as small tumuli of black soil. Kenneth Grahame was right to imagine Ratty and Mole becoming firm friends. After all, they were neighbours.

In Grahame's fictional waterside world the enemy hid in the Wildwood: untrustworthy weasels and stoats. He was almost right, but Ratty's real enemy has proved to be another mustelid – the American mink. Once farmed for their fur, these escaped (or even 'freed' – by animal activists) carnivores have proliferated, and taken the water vole from the River Itchen. These mink are not intimidated by the vole's freshwater habitat, and are the right size to invade their nesting sites with impunity. Nowadays, the Ratty that my father regarded as something of a pest has become a flagship species for conservation. In another part of Hampshire the Meon Valley Project is attempting to stem the decline of our native mammal, which is invariably described as charming, or even charismatic, when once it was merely common. A few years ago I came across a happy family of water voles in a marsh in Suffolk, going about their business single-mindedly crossing and recrossing a dyke, mouths stuffed with weeds. They could have been the same voles I had seen more than half a century earlier. I was taken back to evenings on the River Itchen, and for a few moments became that small boy again, absorbing everything into a sponge-like mind in a world pregnant with possibilities and discoveries. The voles were not swimming in the same dyke a year later and I could not help but fear the worst.

Rivers have always been a metaphor for the passage of time. While I was at school we sang mournfully, and without much comprehension, a famous hymn that included the lines: 'Time, like an ever rolling stream, bears all his sons away'. Ranks of small boys in school uniform with all their shoes polished and their heads full of football failed to appreciate the truth in the familiar lines. The words were just there, like the portraits of previous headmasters in the school hall. When I looked into the limpid waters of the River Lambourn as an aging man I clearly saw the passage of years mirrored in the flow of the stream. I realised the importance of trout in the construction of the person I became. I mused on the passing of time: 'A thousand ages in thy sight are but an evening gone' the hymn continued. Much of my life has been spent contemplating those thousand ages. Once again, I discovered the *suspension* from time that overtakes the conscious mind when watching weeds tossing in the water, and hearing the trickle of currents over the shallows. The river bore my memories away like fragile trout flies.

2

Eggs and After

In 2016 I found a bird's egg on the forest floor of my own small patch of woodland. It has a ragged hole on one side, its contents gobbled up by a thieving magpie. The egg is still as blue as the sky on a cloudless spring day – bright and clear, unsullied with any hint of green or yellow. At its wider end there is a scattering of black dots. Now, lying in the palm of my hand, it weighs almost nothing, but it stirs old memories: I know that the egg was laid by a song thrush. It came from a nest about the size of half a coconut and well concealed amid thick bushes. The outside of the nest would have been woven from grass and moss, but the inner cup would have been an almost perfect hemisphere of dried mud, painstakingly constructed by the parent birds. A clutch of perhaps four eggs lay within the nest, which was perfectly shaped for thrushes to incubate the next generation. As a boy I would have been drawn to the nest after locating the unmistakable music of the song thrush from a tree nearby. The thrush builds its song from short phrases: intensely loud clusters of a few notes are rehearsed perhaps half a dozen times, and then followed by another, different phrase to be repeated in turn. The improvised song can continue for many minutes during the mating season, phrase after phrase, endlessly inventive. These liquid notes used to wake me up in my bedroom in Ealing when glowing crocus flowers were announcing the end of winter with a different kind of display. However, most of all, the discarded eggshell reminds me of my guilt.

When I was a small boy I collected birds' eggs: I was following in my father's footsteps again, though this time not necessarily along a riverbank. My father grew up in a small village outside Worcester called Lower Wick, which has since been swallowed up by the twentieth-century growth of the county town. He was at school during the First World War at Worcester Royal Grammar School. His parents ran a market garden, which was extraordinarily hard work. It was such hard work that Granny Fortey had a stroke in her fortieth year, which left one side of her face paralysed, although she lived on to a considerable age. I never met my grandfather: relentless labour involving wheelbarrows and chaise cloches led to his early death in 1934. My father's was a country life, and it was full of the pursuits that country boys followed in the first half of the twentieth century. The fields and woods were theirs to roam. The River Severn was just down the road, so fishing was a given. Wildlife abounded, and country boys acquired knowledge of birds and plants as naturally as city boys learnt about buses and sweet shops.

The acceptable way to collect a wild bird's egg today.
A predated song thrush egg picked up from my wood.

Egg collecting is often portrayed as sheer vandalism, and sometimes it certainly is just that. The mad lust for a rare egg to complete a collection leads fanatics to illegal and reckless behaviour. My father insisted on strict rules. The first concern was to cause minimum impact on the birds themselves. Only one egg from a clutch was to be taken, as quickly as possible, when the bird was not sitting on the nest. If a nest belonged to a familiar species already collected then it was a case of look, but don't touch. Leave the birds in peace. An egg that was close to hatching could be recognised by holding up the egg to the light – and quickly returned. My father claimed that brooding birds returned to the nest to continue sitting on the four eggs that remained of an original five. I never found reason to doubt this. Then the preparation of the egg was crucial – it should never be allowed to go bad and be wasted. A tiny pinprick was made at each end of the egg to allow it to be blown, or a little circular hole was drilled in the side and the contents puffed out using a delicate blowpipe placed close to the drilling. Finally, a blown egg was placed in a neatly compartmentalised box, with the name and date. There was a collection at home going back to my father's boyhood, logically laid out in several flat wooden boxes; for this young boy it was the subject of endless contemplation. What would have astonished a twenty-first-century birder are some of the species that were included in the collection. The red-backed shrike is now so rare that it sets twitchers all of a twitch. It was so common when my father was young that he saw it on every summer walk and used the nickname 'butcher bird' because of its habit of impaling bugs on thorns of spiky trees. I remember its distinctive, small cream-coloured egg with a band of red-brown spots near one end. When he was a boy it was never a problem for my father to find grey partridge eggs, even though they were hard to discover among the herbage; a clutch may have had a dozen eggs, so that one collected really would not be missed. It has been many years since I encountered a covey of wild partridges scuttling over a cornfield. I *think* I may even remember a corncrake egg in the collection. To see this bird in 2020 and hear its distinctive call the enthusiast has to travel to islands off north-west Scotland. In those neat trays lay irrefutable evidence of what has happened to some of Britain's native birds over the last century.

I confess that I, too, hunted for nests and eggs: the guilt sometimes steals up on me. I also believe that the search was part of my making as a naturalist. I became extraordinarily sensitive to the least movement: the twitch of a leaf, a small bird disappearing into thick cover, the furtive way a warbler would melt into the undergrowth; all pointers to where a nest might be hidden. In the water meadows by the River Lambourn I ran down the home of a reed bunting inside a sedge tussock. I cut my finger on one of the sharp, dangling leaves when I parted their green curtain to find my prize. Discovery was paid for in real blood. I located the tiny domed nest of a willow warbler tucked away in rough grass. I could somehow guess which hole in which tree would house a clutch of a great tit's spotted eggs. This was the atavistic boy, the intuitive hunter, the useful member of the tribe. Like much of what gives rise to science, such concentrated awareness is about making close observations and drawing conclusions from them. The self is forgotten as the seeker puts everything into the senses, sight and sound. The fly fisherman 'reading' the trout's intentions may not be so different. Past experience feeds future predictions: that patch of scrub looks just the place where a thrush might build his nest; this is not a pile of soggy weeds in the stream – it is a dabchick's nest; grey wagtails favour the thick ivy covering on that bridge over a stream. We can outwit nature's deceptions with sharp eyes and a good memory for detail. Nonetheless, some birds *did* fool the canny young hunter. I never discovered where the snipe hid his clutch on the ground in the damp fields, even though I saw the bird in flight many times zigzagging across the marsh; maybe the green and blotchy disguise of the eggs was just too perfect, or perhaps the bird led me astray too convincingly. Anyway, there was already a common snipe's egg in the old collection.

I did add the egg of one species to my father's neatly displayed arrangement: the grasshopper warbler. This is an uncommon and shy little bird, one that ornithologists lump with other 'lbj's' (little brown jobs) that are hard to spot and harder to identify even if you are lucky enough to see one. Like many self-effacing insect-eating warblers, it skulks. Fortunately, it has a very distinctive song, a continuous churr, something like the rasping of the insect that gives the common name to this elusive bird. I must have heard the unusual song and located its general direction, and all those finely honed senses came into play; another of the elevated tussocks made by some variety of tough sedge hid a well-woven nest constructed of strands of grass, from which one egg was removed rapidly – a pretty, rosy-brown speckled affair. I glowed with satisfaction for hours. After the ritual of cleaning, it was placed into the collection in its own little cubicle, safe on a bed of sawdust.

In 1954 collecting birds' eggs became illegal for species other than those that might be considered pests. I have no idea whether my grasshopper warbler find was made shortly before, or shortly after the enactment of the Protection of Birds Act. I do know that collecting that egg would now be a criminal act, but even then I paid a high price for my discovery. Guilt arrived in my childhood and, like inoculation, it never goes away.

My primary school on Pitshanger Lane, on the west side of London in Ealing, was rather a stern building of yellow brick with formal classrooms and wooden desks, and surrounded by a decent slab of tarmac to serve as a playground. It was within walking distance of suburban Ainsdale Road where we lived. I had already been to another school in the more distant suburb of Greenford, about which I recall nothing except a small boy pulling down his pants in class and weeing with enthusiasm, and a teacher becoming exasperated by my inability to learn to tie my own shoelaces, a task I found almost as difficult as managing to work a fishing rod. I think I must have spent most of my time in a kind of amiable fuzz. In due time I arrived at Pitshanger Lane. I assume the teaching was efficient, because I became a good reader at an early age. Playtime was announced by the ringing of several bells echoing in the corridors, and everyone belted outside. A boy known as Piggy Pearson led the gang that everybody wanted to belong to, but I was never invited. While Piggy's gang ran around shouting 'BANG! BANG! YOU'RE DEAD' I preferred to hang out with several girls and two small boys who were prepared to listen to my stories.

One day I was summoned to the headmistress's office. Miss Long was a serious woman soberly dressed whose smile never seemed entirely convincing. Headmistresses then were not meant to be twinkly. There was a complete lack of twinkle when I was instructed to occupy a small chair in front of her. She put on her glasses, framed in black. 'How would you feel', she asked, quietly, 'if you were picked up by a fierce giant and plucked out from your house?' All I recall is the intensity of the stare through her glasses. This was the 'basilisk stare' of P. G. Wodehouse's fearsome aunts, but instead of turning to stone I turned to jelly. I cannot recall my reply, but it must have been along the lines of 'I wouldn't like it, Miss Long.' Her face hardened. '*This* is what you do when you rob a bird's nest! And I know you rob birds' nests.' Somebody had dobbed me in. I could not explain the finer points of taking only one egg and not disturbing the birds that I loved more than anything. I certainly could not explain how the hunt refined my nascent scientific instincts. I had neither the will nor the presence of mind. I was a robber. I was a destroyer of helpless families. I was a giant bringing death and destruction. I knew guilt, and guilt made me writhe and made tears well up. And Miss Long did not even know about the grasshopper warbler.

I think I know who spilled the beans. A boy in my road had junior membership of the Royal Society for the Protection of Birds (RSPB), which was running a campaign against egg robbers. The Society was prominent in getting the law changed. I don't imagine that my young contemporary contacted Miss Long himself, but his father was something important like a bank manager and not inclined to think that there might be more than one way of looking at things. I suspect he picked up the telephone. I arrived home in tears, and my mother was furious. She was more or less convinced of my perfection in all respects, and my distraught little face was more than she could bear. I think there must have been a subsequent scene at North Ealing Junior School in Pitshanger Lane. It was not long before I was taken away to a different school, and I suppose that Miss Long must have been part of the reason, but, as usual, nothing was said at home. When I asked what had happened in later years I was offered a different scenario. I was, said my mother, such a dreamer, that my parents thought I would never pass the 11-plus examination that was the passport to grammar school, and a superior education. So off I went to the junior school of the City of London School for Boys, an excellent private school in the middle of London, which was, of course, a fee-paying *public* school, according to the topsy-turvy nomenclature that is such a speciality of the English. I was nine years old, and I do not remember collecting any bird's egg ever again.

Armed with *The Observer's Book of British Birds* I was now just that – an observer. I have remained so. One of the first birds I spotted near my new school was a black redstart that had moved into a City bombsite left by the Luftwaffe. It was as rare as the grasshopper warbler. I didn't seek out its eggs. Today, I doubt that I retain the skills to discover any small nest deeply hidden among a tangle of brambles. The eight-year-old boy would have got there first.

My father's old egg collection, complete with its grasshopper warbler, simply disappeared. This is some sort of tragedy. It is illegal to trade in British birds' eggs, so it certainly was not 'sold on'. My mother may have accidentally dropped it during one of her many moves. However, the RSPB has encouraged the destruction of some of these collections, and that could well have been its fate. The disappearance of a century-old collection fillets a data source that might have provided a closer look at the time and pace of change affecting British avifauna. My father sampled Worcestershire when most grown-up naturalists were fighting in the trenches. Those records are gone, filed now only in my memory. Was there *really* a corncrake egg? Memories are fallible. Collections are archives of what was there, regardless of moral judgements about the way the specimens were collected. Although the incident of the grasshopper warbler still tweaks a guilty nerve I am not the only scientist to be sceptical about the role of egg collecting in the decline of British birds, just as I don't believe my sister picking marsh orchids had anything to do with their virtual disappearance from the Lambourn Valley.

The garden of our house in Ainsdale Road was full of birds: Ealing was a leafy suburb. My mother often emphasised that *our* road was in postal district W5 rather than W13, an area that she regarded as rather downmarket. She should have known; she was brought up on the smarter side of Ealing, when it lay almost at the edge of London. Ealing Broadway station was the terminus of the District Line in the west and pre-dated other Underground lines that came to link distant villages like Ruislip into the city. I don't suppose birds cared which postal district they were flying through, although I expect the ones in W5 had better vowels. Our garden thronged with house sparrows: Cockney sparrows. The cheerful 'chip chip' of the house sparrow was a perpetual presence around the garden; any morsel dropped by chance was seized in a second: *chip chip* ('ta guv'nor'). We had a large garden for London, with a pond that was forever leaking, and a couple of apple trees that were assailed by bullfinches when the blossom was about to burst in April – they just seemed to delight in destroying the buds. House sparrows and bullfinches have grossly declined since our Ealing days.

Before our family owned Primrose Cottage, fly fishing was conducted from a caravan parked in a field on the Pococks' farm at Woodspeen, just down the road from Boxford, towards Newbury. The two Fortey children ran wild through the farm with the Pocock kids, Roger and Susan. We made camps out of bales of straw, and we jumped from great heights into beds of hay. The Friesian bull was admired – from a safe distance. We spotted huge rats in the cavernous old barn, and they were always heading off furtively somewhere else. There was plenty of mud about, and our knees were usually filthy. We were free to wander over the fields and into the abandoned chalk quarry; cuts and bruises were a small price to pay for unfettered exploration. Nobody thought about hidden perverts. My sister discovered ponies. On the farm, sparrows flitted in small gangs everywhere, arguing amongst themselves. Their scruffy nests were stuffed into the eaves of outbuildings, and Tony the farmer occasionally poked them out, but the sparrows seemed unconcerned. They were back in a week or two. Neater nests of summer swallows perched on the cross beams in the barn. The aerobatics of the swallows provided an unremarked background to our adventures. Just above the farm a tiny single track railway line ran between Newbury and Lambourn: what might have been the smallest station in the world – Woodspeen Halt – was apparently there just to serve the farm. It was little more than a bare platform with a noticeboard. Every two hours a tiny train rattled along the track, but I never saw it pick up a passenger at the Halt. This little spur off the Great Western Railway must have been the least economically profitable line in Britain; a few years later it was axed by Dr Beeching.* Between trains, the track was part of our domain. Thorn bushes had grown up alongside the railway cutting, and in early summer it seemed that every bush included a singing bird. High-pitched recitals of 'a little bit of bread and no cheeeese' progressed in rough sequence along the trackside – the song of the yellowhammer (bunting if you prefer), and it was part of the indispensable music of summer. It was easier to appreciate the song than to identify the yellow face and red-brown back of the bird itself, tucked behind a bush. Small boys are not statisticians, and memory may decorate the past with more than a lick of paint, but I am certain that there were a dozen yellowhammers requesting the passer-by for bread without cheese along the small piece of railway track close to Woodspeen Farm. It was a common bird. In the twenty-first century it has become almost scarce. My attempts to teach my children to recognise the yellowhammer's song – as my father once taught me – have been frustrated, although I still discover a few birds in the Chiltern Hills every year.

* Along with about 5,000 miles of track all over the British Isles. Dr Beeching remains a hate figure to this day.

The decline of this charming omnivore has not been because of the activities of egg collectors; in fact, the downturn in this, and many other species of birds, has happened *since* 1954. The disappearance of their food is surely the cause; after all, there are still plenty of bushes for them to occupy. Once more I recall the difference between two car journeys to the River Itchen half a lifetime apart: the first with my father, windscreen spattered with fat corpses at the end; the second with a body count hardly worth mentioning.

*   *   *

If I felt most alive in the countryside, for much of the time I was just a small boy in a London suburb. Ainsdale Road could be the example of the 1930s expansion of London westwards, part of an extensive development of similar-looking streets that stretched all the way to Park Royal station on the Piccadilly Line. That distinctive Underground station made it into Nikolaus Pevsner's *The Buildings of England* as a pioneering piece of modernism. None of the houses on the estate that included my childhood home would have earned a mention in any guide: they were uniformly unremarkable. Even the most optimistic estate agent could not talk of a wealth of period features. Bay windows at the front, and French windows at the back, a perfunctory bit of exposed timber: that was the extent of their art deco embellishment. Ours was a semi, inseparable from a nearly identical neighbour, like a Siamese twin. A small brick wall separated a modest front garden from the quiet road lined on both sides with similar houses; a garage (with a shared drive) indicated that these were houses for people who could afford a car. The wealthier inhabitants had four bedrooms. Three bedrooms was the norm, like ours: an utterly undistinguished but comfortable home redeemed by a large back garden. Travellers returning by air to Heathrow glimpse uncountable similar estates from far above, where they seem to be draped around the Victorian cores of London's first great expansion. They house millions of people. My mother remembered the houses being built, spreading outwards beyond Perivale to Greenford 'like a rash', she said. When she was a little girl Greenford was a village without electric light. It was an expedition to go there.

We were a tight nuclear family. My mother's devotion to her two children was absolute. My sister Kath is two years my junior, and I was evidently not pleased when she arrived, displacing me from the centre of attention. There has been an undertone of competition ever since, but we played together well in our early years, so long as Kath tagged along behind me. She had her revenge much later. Few relatives visited Ainsdale Road: both of my grandfathers had died young and my mother's mother was to follow suit before I was three. My mother's maiden name was a distinctive one: Wilshin. There had been Wilshins in the old county of Middlesex* for several centuries. The youngest Wilshin and the most regular visitor was my mother's

---

* Ealing used to be in Middlesex, too, always abbreviated to Mddx on the post, until it became part of Greater London in 1965.

younger sister – known to us forever and mysteriously as Auntie Bo – who was always described as the goody-goody, the favourite when they were young, while my mother cast herself as the rebel, the one who got into scrapes. Grandfather Wilshin had several sisters, most of them unmarried, and as the male in the family he had to take responsibility for them when his father died. They were said to have lost their hopes of matrimony in the 1914–18 war. Collectively they were known as the Wilshin Aunts, and they had the full set of names: Jessie, Madge, Gertie, Doris, Marion . . . They always called my mother 'Peggy', which is the name that is engraved upon her christening spoon. The Wilshin Aunts were occasional but memorable visitors to Ainsdale Road, and when they arrived they all carried with them the same smell of lavender bags placed in neat drawers. They tended to wear slightly billowing dresses with flowery designs. They all had terrible sight ('the Wilshin Eyes'). My memory of them is dominated by their glasses, which were as thick as bottle bottoms and made the pupils enormous and shining like an owl's. I had to smile convincingly while they peered through their optical contraptions and commented on the surprising fact that I had grown since their last visit. Auntie Jessie was a sweet soul who sent me a savings stamp worth half a crown on my birthday for as long as she lived. I had to stick the stamp into a book until I was ready to cash in the total. Auntie Marion went mad and was convinced that people were breaking into her house, so she tied strings of cotton all over the windows and doors to catch them out. Both my mother and Auntie Bo had the Wilshin Eyes and wore thick, if more fashionable glasses. My mother was terrified that the curse of the Wilshin Eyes might be passed down to yet another generation, but we Fortey children had perfect sight. This was a fortunate gift for any aspiring observer.

On the Fortey side there was just one visitor. My father had a sister called Anne, who looked after Granny Fortey in Worcestershire. If Aunt Anne arrived in London for a couple of days she would storm out in tears and leave early. Brother and sister did not get on. Return trips to Upton-on-Severn were as bad. My sister and I sat on uncomfortable chairs trying not to move. A clock ticked loudly and slowly, and the daylight was filtered through maidenhair ferns in pots on the window ledge. The only good thing was the garden, on which Anne lavished her attention. I asked the name of a mat-forming creeping plant in one of her flower beds. 'Mind your own business,' replied Aunt Anne, cackling vaguely. It was a joke of sorts. This little (*Soleirolia*) herb's common name really *is* 'mind your own business'. My aunt repeated the joke several times, amused on each occasion. She was briefly married to a Mr Hill who died young. 'Nagged to death,' said my father.

So our small and rather fractious family was not much visited, and my father's fishing friends hardly ever appeared away from the water. Social life centred on my mother's old school friends. She had been sent to a well-known private school, Haberdashers' Aske's School for Girls, near Ealing. Tales about her terrifying headmistress, Miss Sprules, were part of our childhood lore. My grandfather never offered Mother the chance to go on to university after matriculation, something she was angry about for the rest of her life. The 'old girls' remained her best friends. Like 'Uncles' Eddy and Arthur in the fishing-tackle shops several of them became fake 'Aunties'. One of them was to play an important part in my journey to science. There were tennis friends who went to play on Saturdays on the Haberdashers' school tennis courts (a privilege of old girls). I was often taken along and mooched about the school grounds, but never took up a racket, rehearsing my lifetime role as a refugee from sport. There was a gossipy friend who was gooey about a crooner called Johnnie Ray, which my mother found rather absurd. Another friend certainly never became 'Auntie' Barbara. She was, my mother declaimed, 'man mad'. Barbara had lots of lovers, married a postman, and had plenty of children. Although my mother professed disapproval, I detected an undercurrent of admiration for rebellion against the social norms. She 'didn't give a damn'. The only social norm we flouted at 40 Ainsdale Road was painting the exterior woodwork of the house mustard yellow. The other houses on the road were black and white or generic green. The next-door neighbour was vocal in disapproval.

'Auntie' Katharine was mother's best friend. Unlike the other school chums she had a career, as a small-animal vet, so she had the higher education my mother had craved. Her maiden name was Morley-Jones. Her parents lived in one of the finer Victorian houses in the older part of Ealing. They were 'intellectuals' according to my mother, a term which seemed to embody both complexity and mystery to a young mind. I visited their house with my mother on several occasions. Mrs Morley-Jones was a 'bluestocking', I was told, which had me looking for her well-concealed legs. There was a kind of seriousness to the Morley-Jones household that was at the same time intimidating and seductive. Katharine's brother Robert had been sent to Bembridge School on the Isle of Wight, a pioneering place for a liberal education. He was a mathematician of some kind, and rather shy. Mr Morley-Jones was tall and courtly and addressed this small child as if he were an adult. Mrs Morley-Jones had her long hair curled up into two whorls on either side of her head, a little like earmuffs. I never did discover the colour of her stockings. The day that matters most in my story was when Mr Morley-Jones asked if I would like to see his study. I did not know what a study was, precisely. This was the first one I had visited: a small room, rather dark and with one wall lined by cabinets, holding ranks of very thin drawers. Books took up another wall. Our house was far from devoid of books but I had never seen so many together. On a square desk there was a piece of apparatus: a microscope of shining brass above a small revolving stage. Mr Morley-Jones invited me to look down the eyepiece. A bright lamp was switched on to direct light through a slide already mounted on the stage. I squinted to get my eye in the right position, which took a little time. Then I saw the most marvellous things: elliptical plates penetrated by dense pores looking somewhat like colanders, but wonderfully symmetrical. 'Diatoms,' said Mr Morley-Jones, 'that one is about a four-thousandth of an inch long.' I knew only that whatever-it-was was very small and very beautiful. Mr Morley-Jones removed the

slide and replaced it with another: this time it held orna-
mented triangles displaying perfect symmetry. Then came
exactly circular wheels with what looked like spokes
between a wealth of perforations. I understood that the
cabinet behind me with its dozens of drawers must have
held thousands and thousands of these tiny objects. Nature
operated at all scales, not just as birds and fishes and trees.
You could go down and down, smaller and smaller, and
there would still be more to see. Mr Morley-Jones had
selected these tiny algae from the infinity of living things
to make them his own. Those ranks of books must have
been part of his quest for knowledge. The *Observer's Book
of Everything* could not exist; there was simply too much
to know. That microscope was a magic portal into this
other realm.

'Auntie' Katharine also helped to complete our little
family. She married a farmer – 'Uncle' Phil – a countryman
with an accent as rich as well-rotted compost. They lived
on a spectacularly unmodernised farm deep in the country
near Petworth, where instead of having children Katharine
bred Shetland sheepdogs for show. The dogs were fed on
ghastly bits of sheep that hung up in their cavernous
scullery. She worked for the People's Dispensary for Sick
Animals, and sad rescue dogs were part of the job. Several
of them became our pets. A small fox terrier named Sue
was the most beloved member of our family. She arrived
as a shivering wreck, and little by little lost her fear to
become totally devoted to us. She loved coming to
Woodspeen Farm, where she would tackle rats almost her
own size. No small dog has ever had more affection heaped

upon it. We knew nothing of her history, but one day we discovered something by mischance. My sister and I had realised that by blowing up a brown paper bag and twisting the top to stop the air escaping we could generate a satis-factory bang by bursting the bag. Once, when we started this process with Sue in the room, the little dog went berserk with terror, yelping and scrabbling at the door to escape. She must have been tortured that way by her previous owner, and repeatedly. At that moment I under-stood what cruelty was, and it was not taking an egg from a nest. It was several days before Sue could forgive us.

Coronation Day, 1953, was not only a big day in the life of the nation it was a big day in the family, because a television set was purchased. It was a small screen in a big box, and those without television squatted down rapt before it, including Auntie Bo and her dog, a big chow-chow. The dog and I soon became bored. I offered to take the chow for a walk on its lead, and my offer was accepted with a wave as the royal coach made its encrusted way to Westminster Abbey. The problem turned out to be that instead of me walking the chow, the chow walked me. It tugged me up Ainsdale Road and into Birkdale Road beyond, and it dragged me onwards past Sandall Close, the site of the maisonette I was born in, where a doodlebug had demolished a neighbouring house during the war; I could not stop the dog marching me ever onwards into unfamiliar territory. Fortunately, there must have been virtu-ally nothing on the roads (*everyone* was watching the Coronation) because dog and boy somehow breached the Western Avenue, with me hanging grimly on to the lead, and onwards into the Haymills Estate. I was now completely lost and frightened, and probably weeping. The only republican in Ealing W5 was cutting his hedge and realised something was wrong as I shot past. I could just about say where I lived: '40 Ainsdale Road,' I gasped, between sobs. I arrived home in a police car. That is all I recall about the day our monarch came to the throne; the Queen who still reigns as I write these words. Nor can I remember the name of Auntie Bo's chow.

In these early days BBC television programmes were not broadcast all day. They were interrupted by restful shots ('intervals' they were called, in black and white of course) of potter's wheels turning or of wind playing over reed beds. When nothing at all was showing there was a curious, semi-abstract, but entirely motionless image called Test Card C. I assume it was supposed to demonstrate the range of grey shades the TV could encompass. I loved the new television, and if I had had the chance I would doubtless have gawped at it all day. As it was, after about an hour of sitting in front of Test Card C waiting for something to appear, I gave up. That was an excellent thing to happen in Ealing, W5. I was forced to *do* something, or read. I invented a game of my own based on the famous London Underground map. I would draw up new lines carefully on the map connecting Piccadilly and Bakerloo with District and Metropolitan Lines in ways undreamed of by town planners. On my new line a passenger could travel from Upminster to Uxbridge, by way of Bounds Green, East Finchley and Harrow & Wealdstone, in a great loop circumscribing North London. Or he might wish to go from New Cross Gate to Hounslow by way of Lambeth North and Wimbledon Park in the south. The new route might be called the Lakerboo Line or the Dillypically Line. For some reason, few of my passengers wanted to go into the middle of London. My idiosyncratic invention may, or may not, have something to do with my subsequent development. Bertrand Russell wrote an essay *In Praise of Idleness* arguing that lack of frenetic activity, or even boredom, may actually be a stimulant for the creative instincts, and this was about the best I could do. I began to devour books with complete lack of discrimination: *Just William* and *Biggles* and *The Child's Garden of Verses* and *The Water Babies* and books from the grown-ups' bookshelf. The oldest book was very old indeed, so old that 's' was printed as 'f'. It was an illustrated edition of *Aesop's Fables* ('The Town Moufe and the Country Moufe' was included) and I marvelled at its antiquity. It was also the smallest of any of our books, almost a miniature, and the simple, tiny illustrations – woodcuts possibly – headed up a tale that invariably finished with a moral: *hard work defeateth sloth* (or something like that). I don't recall *the Devil maketh work for idle hands* but Bertrand Russell would certainly have disagreed, which is the problem with morals in general, unless you are Miss Long. From the grown-ups' shelf, and quite high up, was *The Psychology of Insanity* by Bernard Hart, full of intriguing words like 'psychosis' and 'Oedipus'. Much later, I learned that my father had had some kind of nervous breakdown before the war: Hart's book must have been purchased at a time of crisis.

The Second World War left behind one memorial at Ainsdale Road. At the end of the garden was a miniature

Anderson Shelter. It was half a tunnel roofed with over-lapping sheets of corrugated iron. When the bombs arrived the family was supposed to crawl into the shelter until it was safe to go out again. In its redundant state it made an ideal 'camp' for a small boy. I retreated there on uneventful days to make what I termed inventions. They were combi-nations of bits of wire, old springs, string and whatever else I could find. I may have invented a time machine by accident on a wet afternoon, but if so I have forgotten how it was done. The inventions probably had more to do with Rowland Emett's mechanical marvels and crazy machines that I had seen at the Festival of Britain in 1951. Let us be kind, and call the inventions works of art.

For two years I took the District Line train from Ealing Broadway to Blackfriars to the City of London School. Very little about that journey has changed on that old line, although I do not see many ten-year-old boys travelling alone, as I did. The school was on Victoria Embankment in the middle of the City by the River Thames close to Blackfriars Bridge, and boasted a rather grand entrance, complete with columns. It was decidedly impressive after Pitshanger Lane. A splendid hall housed assembly for the whole school in the morning, when Dr Barton, the head-master, addressed us from the podium. The little boys were at the front. The many achievements of the bigger boys were announced. It was all rather awe-inspiring. I encoun-tered a foreign language for the first time, when Monsieur Field encouraged us in oral French. '*Ou est le plafond?*' '*Je m'appelle Fortey*' – that kind of thing. Mr Lewis, the form master, taught almost everything else. My fellow pupils were from wealthier families. I was once invited to a birthday party at a large house in North London, where there was a conjurer who took out coloured ribbons endlessly from an empty hat, and made coins disappear. This sort of thing did not happen in Ealing, even in W5.

One day I did not go to the City of London School. Instead, I went back to Pitshanger Lane to take a test called the 11-plus. Miss Long was there, and even smiled wanly and said it was good to see me back in her school. I found a place in a room full of other children sitting at desks and went through various interesting exercises on paper to decide which number came next in a series of numbers, or what pattern fitted into which template, and then some fairly simple stuff with words. I did not realise the result was important. I passed the exam quite comfortably. I think both my parents were surprised to discover that I actually had a brain. They were probably relieved in equal measure that they could stop paying school fees when I was admitted to Ealing Grammar School for Boys. I was sad to leave the school at Blackfriars; Mr Lewis said how sorry he was to lose me from the class. After my last day I looked so bereft on the way back to Ealing Broadway station that a lady on the Tube gave me two shillings. I must have that kind of face.

*   *   *

Nothing was said, but the family must have been getting more affluent through my early life – the television set was an early indication. The post-war years were increasingly optimistic, and Prime Minister Harold Macmillan informed us all in 1957 that we had 'never had it so good'. Anglers had more money in their pockets to buy superior rods and more luridly coloured gentles. Machines freed women from boring chores like sweeping and washing. Every time we went fishing we passed the splendid art deco Hoover factory in Perivale, on the Great West Road. The building – a palace, more like – carried the legend: 'It beats as it sweeps as it cleans' in great big letters, which description of their eponymous appliance neatly encapsulated domestic liberation. Our first family vehicle was a black affair with a small, oval back window, an automobile in which George Raft might have fled a heist in a dozen of his movies. A series of shapely Austin cars followed, each a little larger and faster than the last. At some stage a second car for my mother appeared, an ancient 1930s Morris with the registration number beginning ALD, and immediately nicknamed 'Oldie'. The old crock gasped and wheezed westwards to take us kids to Ruislip Lido on hot summer afternoons, where we could splash around and get cool and look for dragonflies. Then its 'big end' went, and the car expired on the A40 uttering a series of spectacular thumps until it ground to a halt and the dog jumped out of the window. Undeterred, my mother bought a limousine for a song, an Armstrong Siddeley Sapphire coupé, no less, with real leather seats, and a top that folded back in the sunshine. It could accommodate all our young friends on the way to the lido. I believe these were the best days of my mother's life: children who adored her, driving an Armstrong Siddeley, which she would have described as 'mad' in an approving way.

There were several caravans in succession on Woodspeen Farm, ranging from spartan to quite luxurious. The first caravan was impossibly small and there is no record of its title; the next one was called the Eccles Elite and had tables that turned into beds and a kitchen that folded out of the walls; the largest, the Stanmore Deluxe, was almost comfortable. From anonymous to elite to deluxe – what could better summarise material progression? When the rain pummelled down on the roof of the Stanmore my sister and I played board games, or headed for the barn to find the Pococks.

My parents were not generally extravagant. They were sparse drinkers: wine only appeared at Christmas dinner, and then it was a sweet Sauternes ill suited to the main course. The main indulgence was my father's smoking. He could have smoked for the national team. The favoured brand was Player's Navy Cut – untipped. There was a picture of a jolly jack tar (appropriate, it turns out, for all the wrong reasons) on the packet, and the motto 'It's the tobacco that counts' appeared on the flap of the packet when it was opened. He would give each cigarette a short tap on the packet before lighting up yet again. His fingers carried the stubborn yellow stain of the nicotine addict. My sister and I must have been surrounded by a fog of smoke throughout our early years, even in the car. The only object I still own that bears testimony to that habit is a battered brass ashtray advertising Bass in Bottle that was alleged to have been taken from the NAAFI during the war. It has four depressions to hold smouldering gaspers, enough for about an hour. The consummate angler had his faults, and what others might there be?

At about the same time I realised that my mother was an awful cook. It was not a skill that could be honed during years of rationing, but even when bananas, brisket and butter were freely available she was still stuck in an era before cooking was regarded as a necessary skill. My sister tells me that she cannot recall eating at all; I remember only that Kath subsisted almost entirely on potatoes (Mother even dubbed her 'Little Miss Murphy'). I have some recollections of boiled mince with tinned peach slices to follow. The only thing of culinary excellence we consumed was father's trout – and *he* cooked them. We even looked forward to the school dinners that most children despised: there were delicious things like jam roly poly with bright yellow artificial custard on top. The real excitement at home was Sunday lunch. Like many middle-class families, we all had to come together for a weekly meal cooked by a beneficent mother, a ritual to which the dining room was dedicated. It was sometimes a terrifying experience. Joints of beef would be roasted until they halved in size and became almost impossible to carve; Yorkshire pudding somehow carbonised at the rim while still being soggy at the centre. Greens were mercilessly boiled. We were all grateful to whoever invented gravy powder (a Mr Bisto, I presume) to give us something with which to lubricate the plate. The excitement came with the dessert, especially if it was apple pie. I do not know how she managed it, but my mother's pastry was some kind of hard and brittle ceramic. The pie arrived at the table in an enamel dish looking like an archaeological object unearthed from Thebes. The usual approaches to crust simply failed; knives bounced off it. Eventually a cleaver of sufficient gravitas was applied with great force, whereupon the pie crust suddenly shattered sending pieces flying around the dining room. Nothing was safe: bits bounced off the mirror with a 'ping!' The lampshade trembled under impact. The dog hid under the sideboard. It was remarkable, but the sliced apples within the pie had managed to stay hard. 'Now,' said my mother, 'who's for seconds?'

My mother's cooking never improved. Late in life she did appreciate that she was required to entertain, and that cooking might be involved. She was delighted when she discovered a gourmet chicken recipe that required her to tip a tin of Campbell's condensed mushroom soup over some chicken legs before putting the dish in the oven to cook. Her elderly friends were impressed (according to my mother's account): 'Margaret,' they said, '*so* delicious! *Do* tell us your secret.' She was – as she would have said – tickled pink by her cunning.

After Sunday lunch it was time for *The Brains Trust* on the television. It would be hard to imagine such a serious display of erudition being regarded as entertainment today. If we would have preferred to watch *Roy Rogers* we did not say so, and anyway it was not an option. We all sat in front of the 'box' as if in a classroom before a respected teacher. Questions were sent in by viewers and debated by three panellists, all of them noted savants of the day. Among those who appeared were Julian Huxley, probably the leading British biologist in the middle of the last century, Jacob Bronowski a renowned scientist and thinker,

C. E. M. Joad the first television philosopher, A. J. Ayer a more distinguished academic one (see Chapter 8), and Marghanita Laski, one of the cleverest women in the United Kingdom. There were no camera tricks, just verbal banter, and a good quantity of smoke from pipes. Nor were the questions of the 'What is your favourite book, and why?' variety, instead, they cut to the chase on the meaning of existence. There were questions like: 'Can we ever know ultimate truth by means of science?' C. E. M. Joad was renowned for replying: 'It all depends what you mean by . . . ultimate truth.' 'Is music the most perfect of the arts?' came the question. 'It all depends on what you mean by . . . perfect,' came Joad's response. Nonetheless, the quality of intercourse was very elevated (so elevated, indeed, that the word 'intercourse' could be used in its original sense) and I recognised words that I had first seen in *The Psychology of Insanity* – like 'archetype', 'subli-mation' and 'catharsis'. Words, I began to understand, were legion, multifarious, prolific – and powerful. I needed to get on top of words. I also appreciated the knowledge and wisdom that invariably typified Huxley's replies. I did not know then that he was a member of a famous scien-tific clan that was founded by T. H. Huxley, defender of Charles Darwin and evolution. It would not be many years before I encountered Julian Huxley again in the scientific literature. I understood now what my mother had meant when she described Mr Morley-Jones as an intellectual – he was somebody that would have felt comfortable with confronting difficult questions. He appreciated scholar-ship, and admired reason. Sadly, on a black-and-white

television I could not tell whether or not Marghanita
Laski's stockings were blue.

*   *   *

The butterflies that thronged on the railway embankment
above Woodspeen Farm intrigued me. I could have identified them from *The Observer's Book of British Butterflies*
readily enough, but I felt compelled to make a collection.
I had admired prepared butterfly specimens in old cabinets
in museums with their wings displayed neatly flat and
four-square. Few butterflies come like that naturally: they
need to be pinned out for display, which requires a measure
of legerdemain. I obtained a simple butterfly net and
captured the poor creatures without too much difficulty.
The main problem was extracting them from the net
without destroying the beauty of the wings, the very thing
that attracted me in the first place. Practice improved the
delicacy of my handling, but then my victims had to be
turned into a jar and killed before the pinning-out process.
I obtained a pungent liquid called carbon tetrachloride (it
was used for dry-cleaning as a solvent), and soaked blotting
paper in the jar in the lethal fluid. When the butterfly was
added to the jar the lid was screwed into place and the
insect briefly fluttered, then stalled, and twitched before
expiring. At this stage its wings could be moved into position and pinned into the mounting paper. I did not persist
with this pastime for very long, though long enough to
learn the common southern English butterflies. I made a
small, and not very skilled, collection of spread-eagled

species. Something about the twitchy ending of a fragile life in a jar gave me pause. I knew I could never become an obsessive lepidopterist. I am not proud of this period of slaughter, but it has left a legacy: my memories of abundance. Small tortoiseshell butterflies, whose caterpillars

My sister's sketch of Sue, our beloved fox terrier.

feed on the common nettle, often made orange clouds at the edges of the fields. Red admirals were nearly as common, unmistakable with their stripes of scarlet command. I captured wall browns with no difficulty. These familiar butterflies are not as common as they once were. The culling of the odd adult for a collection is an irrelevance in the prosperity of the species. It may be the birds' story all over again. I am sure that nettles are *more* abundant than they were in my Woodspeen Farm days. Nitrogenous fertilisers have soaked nearly every ditch and hedgebank and encouraged nettles at the expense of a richer suite of wildflowers. Small tortoiseshells should be everywhere. Instead, the populations of small tortoiseshells have fallen by three-quarters in thirty years. The word 'baffling' has been used in official reports.

3

The Chemi-shed

A few items have survived from my time as a boy chemist: just a selection of my old glassware like flasks, bottles and a funnel. The bottles have serious, ground glass stoppers, and they were designed to contain equally serious chemicals. They hold rather less than half a pint of liquid, and any labels they once carried have long since fallen away. The flasks include one with a triangular section, designed to allow the user to swill around potions without slopping them on to his sleeves. Its special glass ensures that boiling liquids will not shatter the vessel. The small glass funnel has gently fluted sides, and readily takes the folded pleats of a filter paper; cloudy liquids poured into it emerge clear into the flask below, leaving behind insoluble residues on the paper. I must have used the funnel dozens of times. How these items came through the vicissitudes of my history is a mystery. Like the famous trout, they survived my mother's frequent moves; they turned up without biographical notes in a box at my sister's house at the beginning of the twenty-first century. The more complex pieces of apparatus, like my retorts or my Liebig condenser, must have been shattered when they were jolted too vigorously in some removal van. No trace of a rank of little screw-cap bottles has survived: they contained an array of chemicals for my experiments, substances once so treasured and so pregnant with possibilities. Coloured crystals were then carefully stored for future use in brown bottles, with their chemical formulae neatly written on the labels; mineral salts: pale green of iron, blue of copper, or purplish cobalt, all gone. Acids, alkalis, sulphates and chlorides must have been poured away in an abandoned chaos of elemental confusion when my chemistry shed was finally emptied. The very pipes carrying them away would have been corroded. There were no explosions that I recall.

Chemistry began with a Christmas present, a chemistry set in a cardboard box with a smiling boy in short trousers on the lid doing something interesting with test tubes. There was a line of similar tubes inside the box and some of them included chemicals for my first experiments, designed specially for a budding scientist. I suspect that several of these chemicals would now be proscribed under health and safety legislation, as they included poisonous substances like copper sulphate, but I never felt compelled to drink a solution of anything to find out what might happen. A small spirit lamp was included to heat solutions in a chosen test tube, which had a lip that could be easily grasped by a simple set of tongs. The most exciting single item was a length of magnesium ribbon that came in a circular pillbox. When the ribbon was held in the tongs and heated in the flame of the lamp it burst into a white blaze of extraordinary incandescence that burned slowly to consume the silvery metal. All that remained was a white wisp. My delight in this simple experiment was unbounded. It demonstrated that one thing could transform into another. The world was not fixed, like school rules. I learned from helpful notes lying in the box that the proper word for what I had seen was 'combustion'. Many ordinary things burned in all manner of ways, like my mother's Yorkshire pudding, but matter itself is transmogrified by combustion. The white residue was still in some way magnesium, but

A funnel and a few reagent bottles –
all that survives from my chemical boyhood.

magnesium mutated, and the process of mutation gave out heat and light. If the transformation happened in air, then the only place from which the mutating agent could be drawn was from the air itself. Air could not be empty; 'vanishing into thin air' was not possible because the very air had substance. I understood that oxygen was necessary at the same moment I realised it existed. Like the mayfly, the inanimate world could transform from one state to another: existence was a state of flux.

There were other transformers waiting in my original chemistry set. Ammonium dichromate was a mass of orange crystals; if placed into a small paper pyramid and then ignited at the tip – sparks flew. A spitting, glowing combustion made a convincing replica of an erupting volcano; it even hissed in a satisfactory way. The most extraordinary feature of this transformation was that the pile of 'volcanic' debris was many times the size of the original dichromate. The orange crystals puffed up into a huge pile of a completely different dull greenish chemical (chromium sesquioxide I discovered later). The vigour of the reaction somehow suggested that the dichromate sought actively to change, as if motivated by desire rather than chemical inevitability. To young eyes this was as much magic as science: was there no end to possible transformations? There was even disappearance – the conjurer's stock in trade. A few yellow sulphur crystals placed on a spoon in a flame glowed and were gone. But they could not have vanished into thin air, because there *was* no thin air. A sharp assault on my nostrils made me guess what had happened: no elemental prestidigitation, no chemical hocus-pocus. The

combusted sulphur became a gas – not so much disappearing into thin air as thickening it with its own element.
I recognised the same smell again when I went to Spain
as a teenager and travelled on the old coal-fired steam
engines. My eyes smarted, and my throat tightened into a
cough. Sulphur hidden in poor quality coal was combusted
into the atmosphere as sulphur dioxide. It was poison. It
lurked in smog. It killed lichens that were immune to
drought and starvation and the wrath of God.

There was a particular logic to chemistry that I was just
beginning to understand, but something equally stirred my
imagination. The idea of metamorphosis is embedded in
so many mythologies and in so many cultures that it must
be rooted deep into the human mind. Greek gods trans-
forming themselves into animals, trees embodying spirits,
spirits then turned to stone, even transubstantiation; so
many of the colourful costumes of religion are concerned
with one thing becoming another, animate and inanimate.
What Jung termed the collective unconscious, where such
transformations are part of the psychic landscape, may be
a legacy inherited from a common ancestor, an ancestor
so deep in geological time that the boundaries between
man and the natural world – between the animal, vegetable
and mineral – were not taken for granted. They were
blurred by superstition and need. The emotional appeal of
chemical transformation may run keenly alongside a
rational desire to find out how matter works. We anatomise
substance, or even recognise atoms, spurred on by a
primeval fascination with metamorphosis.

I was given the book some time after the chemistry

set. It was called something like *Chemistry Experiments at Home* and had a brown cover (dust jacket unknown); the author's identity has escaped me, although the name of 'Robinson' popped into my head briefly and for no good reason, so Robinson it shall be.* I owe Robinson a big debt. He identified chemicals that could (then) be obtained for domestic purposes: sodium carbonate as a cleaner, chlorinated bleach, silver nitrate and 'hypo' for photographic development, sodium chlorate weed killer (now banned) potassium nitrate (banned), ammonia (still a cleaner), acetic acid (vinegar), potassium hydroxide (lye) and so on; they augmented my chemical arsenal considerably. Robinson told me how to grow giant copper sulphate crystals by suspending a very small one on a thread of cotton in a supersaturated solution of the same chemical. I loved the old names of some compounds, which went back to the days of the alchemists. Copper sulphate was 'blue vitriol' and sulphuric acid 'oil of vitriol' – which immediately explained the adjective 'vitriolic'. Potassium nitrate was once termed saltpetre – an ingredient of explosives, as I was to learn – but in those days I could buy a pound of it with no trouble. Robinson detailed a range of experiments using such easily accessible chemicals. One of the first I sought out was sodium silicate, sold as 'water glass', which was formerly used to preserve eggs. It came in a tin, and was a kind of

---

* When this book was in press my wife discovered the likely identity of 'Robinson': *Chemistry Experiments at Home for Boys and Girls* by H. L. Heys.

thick, colourless syrup. After it was diluted and the solution poured into a jar Robinson's young chemist was invited to drop in crystals of copper sulphate, iron sulphate and similar metallic 'salts'. I followed his instructions and over a day or two the most fantastical towers grew up into the solution from the crystals, composed of silicate compounds of the 'seeds' in the jar. They bore the colours of their elements: blue for copper, green for iron. The growths were frayed, carbuncular and delicate, extravagant as stalagmites. They resembled the phantasmagorical columns portrayed in Max Ernst's surrealist paintings made a few years before I was born. Once again, it was transformation, chemical creation, molecular magic. More prosaically, Robinson told me how to make a detergent from horse chestnuts. By boiling up the brown nuts I released a satisfactory foamy solution; the same chemical became the 'secret' ingredient of Badedas bath bubbles.

The book also taught me chemical equations, through friendly words at first, rather than chemical symbols. That first flare of magnesium was the simplest possible equation:

$$magnesium + oxygen = magnesium\ oxide$$

(wrote Robinson). Two elements combined to make a compound molecule, the white residue. I understood that equations can go both ways, but to recover the magnesium would mean reversing all that heat and light – energy – that had been expended during its brilliant transformation. It was as if magnesium *wanted* to marry up with oxygen and was reluctant to part with it again. When I learned much later that elements formed *bonds* it seemed just the right word.

Robinson led me easily through a slightly more complex case history using only the most everyday materials. I retrieved a lump of chalk picked up on one of the family fishing trips and roasted it on the gas stove. Eventually, it started to crumble. It had become lime. The relevant equation was simple enough:

chalk (calcium carbonate) − carbon dioxide
= lime (calcium oxide)

Carbon dioxide (being a gas) was given off to the atmosphere to contribute minutely to global warming, and only lime remained behind. The next phase in the experiment was to shake the powdered lime in a test tube with cold water. I noticed that the tube became quite warm for a while. There was little to see, but some of the white powder had dissolved as it reacted with the water:

lime (calcium oxide) + water
= calcium hydroxide (limewater)

The old way of describing this process was 'slaking' the lime, as if it were thirsty − almost as if it were seeking to bond with the water to quench some molecular desire. Calcium hydroxide is still known as 'slaked lime'. Now I had to pour off the limewater into another tube, leaving behind any undissolved lime. The final phase of the experiment was to put a straw into the clear limewater and blow.

As my breath bubbled into the clear liquor within a few seconds the limewater had turned all milky. The final equation explained it all:

> limewater (calcium hydroxide)
> + carbon dioxide (my breath)
> = chalk (calcium carbonate) + water

Chalk is insoluble, so it precipitated to form the white clouds in the test tube. The whole experiment had come full circle: from chalk, and back to chalk. I used my little funnel to filter off the chalky white calcium carbonate on to a folded paper. Pure water passed through into the flask. This experiment could go around and back (unlike burning magnesium) because the energy used to drive off the carbon dioxide from the chalk sample was 'recovered' as heat when the lime was slaked. As for the chemical formulae used to describe scientifically what had happened – I soon learned the advantages of their brevity and concision. That last equation is written simply:

$$Ca(OH)_2 + CO_2 = CaCO_3 + H_2O$$

and provided a way of recording the symbols for the elements involved and making sure that there were the same numbers of atoms on both sides of the equation. I soon learned to feel comfortable with chemical symbols, but the language of yearning and metamorphosis infused my chemistry for a long time. I was not yet the pure, dispassionate scientist.

Something had to happen to accommodate my growing collection of chemicals. At the end of the garden in Ainsdale Road was a small, weathertight wooden shed, and it became my first laboratory. I soon had a bench to work at and shelves at the back to store what my mother insisted on calling my 'potions'. A spirit lamp gave way to a Calor gas Bunsen burner. Every birthday brought another piece of 'kit' to what was first known as the 'chemistry shed' and then abbreviated to the 'chemi-shed'. I went in for pipettes, and accurately calibrated flasks. An old spring balance was not exactly a precision instrument but it was adequate for schoolboy recipes. I could distill and I could condense and I could evaporate. I had a pestle and mortar to grind stuff and a strange metallic retort that could have come from Paracelsus' cellar. I could disappear into the chemi-shed for hours and my parents thought I was gainfully occupied. They had very little idea of what I was up to. I admit that stinks and bangs were quite high on my agenda. My neglect of safety issues was absolute. I made hydrogen sulphide, the gas that smells of rotten eggs (I had better not say how I made it) and I rejoiced in its unpleasantness. Only five parts per million are needed to produce its characteristic pong, and that is probably just as well because it is mightily toxic. At high concentrations it kills rather rapidly (it has even been used to commit suicide). The very idea of a boy, not yet in his teens, wafting around carrying beakers emitting this virulent stuff would result in parental panic today. As for my own detonations, gunpowder was easy. I had plenty of sulphur and saltpetre, and in my mortar I ground powdered charcoal from larger pieces that lurked forgotten under the bonfire heap. Black gunpowder was just a combination of these three ingredients, allowing me scope for experimentation in different proportions. I did manage several satisfactory explosions. I tried mixing sodium chlorate and sugar for another explosive recipe, but I did not have so much success with this combination, which tended to flame vigorously without making much of a bang. With the addition of a few iron filings the mixture became the basis of some crude fireworks and my sister insists I got her to hold some of my fizzing, flaming prototypes in her hand, while I watched from a safe distance.

I had a secret source that helped me extend my range of chemicals. Next door but one to Tooke's, 614 Fulham Road, was a dispensing chemist. Mr Ehrbar was the pharmacist, and owned the slightly dowdy premises. Big blue phials of copper sulphate were the only gesture towards decoration. He was a quiet man with a central European accent, who wore a white laboratory coat, and had a world-weary air about him, as if he were an émigré who had seen too much. He befriended the young chemical enthusiast. This is where I filled my glass-stoppered bottle with oil of vitriol. I am sure that it was illegal to sell a young boy concentrated sulphuric acid but he did not seem to mind. I like to think he wanted to encourage me along a chemical pathway. Even he would probably have drawn the line at giving me arsenic. Oliver Sacks had an uncle who supplied him with just about any chemical element he required. Sacks described the stories of his boyhood investigations in *Uncle Tungsten* and he seemed to be able to recall every experiment he had ever conducted. My own agenda was far more random. I certainly remember asking Mr Ehrbar for picric acid (he obliged) for no better reason than I wanted to synthesise the related chemical potassium picrate – which explodes very satisfactorily upon percussion (and it could be used as a fuse). I gradually worked through the chemicals that Mr Ehrbar hid behind the scenes in his back room. I suspect that many of them were as obsolete for medical treatment as mandrake root, so some of his storage jars may not have been opened for years until I came along. My chemi-shed shelves filled up in a satisfying way, from aluminium to zinc, and I knew by heart the appearance and even *feel* of many of Mr Ehrbar's contributions. I learned the litany of the chemical elements, the basic building blocks of my collection. I began to understand what transformations were within my grasp, and those I would never achieve from my own bench.

It was as well that I was not a psychopath. A near contemporary of mine was Graham Young, 'the teacup poisoner'. He, too, became enamoured of chemistry from an early age, and started his career with a chemistry set, but rather than stinks and bangs, he became skilled in the sly arts of the poisoner. When he was thirteen, while I was closeted in the chemi-shed, Young had already tried out cocktails of dire chemicals on his school friends, and then pursued his grisly studies further on his own family. He obtained those elemental poisons – antimony, arsenic, thallium – that even Mr Ehrbar would never dispense. He knew all about alkaloids, the deathly secrets of deadly plants. After several planned trials, he lethally poisoned his stepmother, watching her final agony with dispassionate interest. He was sent to Broadmoor at the age of fourteen ('the hospital for the criminally insane' as it was then known), its youngest inmate since 1885. While incarcerated in Broadmoor he was consulted on other cases, because of his extensive knowledge of toxicology. He was released when he was just twenty-three, and lost no time in returning to Willesden to gloat over his previous crimes. He must have been lurking just around the corner from Harding's fishing-tackle shop, as my father constructed his split-cane rods. Within a few years Young had killed two further victims, still in thrall to his macabre chemical obsessions. He died in gaol in 1990 at the age of forty-two. He must have looked upon his human prey much as I looked upon butterflies quivering before death in a killing jar; but I had more sympathy for the pain of those few grams of fragile living arthropod than he had for his own family. It is evidently not difficult for human beings to view their fellows as no more than 'specimens' – this is the tragic tale of every genocide – but civilised society in peacetime regards such judgements with horror, and describes them as aberrations. Graham Young was the ultimate aberration.

Many years later when I began to write in earnest I wondered to what extent writers have to become dispassionate observers of their fellow beings. Do they treat people as specimens? Observations on personalities or idiosyncrasies are not altogether dissimilar to those made on morphology by a lepidopterist or a palaeontologist. Writers need to pin the features of their characters accurately down upon the page: they observe from the sidelines in the role of 'the cat in the corner', as John Updike described it. A certain measure of disengagement is part of the process. The peculiar chemistry of writing resides in the metamorphosis of observations from life into written character: an irreversible reaction, another rearrangement of the molecules.

I did pass the Graham Young Test. That is, I had poisons in the chemi-shed which could have obliterated the entire population of Ainsdale Road, and I chose not to do so (even the banker whom I thought contacted Miss Long). My lethal gallery was courtesy of another book, or rather a series of books, an antiquated but comprehensive encyclopedia of knowledge in blue covers that we had acquired in a 'job lot' at auction while Primrose Cottage was being furnished. I believe it dated from the years after the First World War. Like most encyclopedias the entries were alphabetical: Aesop to Arsenic was but a small step. The chemical entries were a poisoner's delight. When I looked up hemlock, I found a cross reference to 'coniine'. I followed through to its separate entry and found a paragraph much like: 'Coniine, a poisonous alkaloid with the chemical formula $C_8H_{17}N$ which induces paralysis and death within a short while after consumption. It can be obtained from hemlock (*Conium maculatum*) by boiling the young leaves and reducing the liquor and then applying alcohol extraction. The death of Socrates is attributed to hemlock poisoning.' I could recognise hemlock without much trouble – it grew by the River Brent in Pitshanger Park, a herb taller than I was, with purple-blotched stems and feathery leaves. The encyclopedia entry was a gauntlet thrown down to a young chemist, and before long I had some colourless coniine crystals in my evaporating dish to put into a little bottle with an appropriate label. Next entry: 'Digitalin formula $C_{36} H_{50} O_{14}$, a poisonous alkaloid inducing heart failure in excessive doses, and obtained from foxgloves (*Digitalis purpurea*) by . . .' Foxgloves were one of the few plants that grew in our garden with abandon, clearly to challenge my chemical expertise. Or 'Monkshood (see Aconitine)', a blue flower in our herbaceous border and a recipe for a quick death . . . I suppose that my motivation for extracting the deadly essences from these herbs was akin to that of the gun collector who never intends to fire a shot. They were a secret cache, a testament to expertise, a guilty but delicious 'what if?'

If that old encyclopedia of arcana was my 'dark web', then my Wikipedia was another set of encyclopedias that I leafed through almost as soon as I could read fluently. Arthur Mee's *Children's Encyclopedia* was a cornucopia of facts in ten volumes. The first version of this work was published in parts before the First World War, but the one we had at home was a different and later version published shortly before the global conflict that was to follow within a few decades. Nonetheless, the maps of the world that decorated the frontispiece were mostly coloured pink – the British Empire still intact, and appropriately tinted Africa, Canada, India and Australia, and many more patches of the world besides. The values inculcated by Arthur Mee were of his time, and must have sunk deep into my pores, to be sweated out slowly during future decades. The

Empire was – in the supercilious classification of *1066 and All That* – a Good Thing. There were photographs of peoples from around the (pink) world: fine warriors from Africa holding spears and shields, Australian Aborigines waving tubers, Maoris splendidly tattooed, Indian maharajahs as grand as could be on elephants draped in finery. There was, I believe, no overt racism – we were invited to marvel at the rich variety of the peoples of Empire – but everything was underlain by an assumption that we British were beneficent purveyors of the right stuff. The English language allowed Hottentot and sepoy alike to rejoice in the magnificent writing of William Shakespeare. British justice was the model for all courts; fair play was an export that came as a free bonus from colonialism. There are revisionist historians who purvey somewhat similar views even today, and I suppose I must have uncritically accepted imperialist notions when I was young. As for the geography of these diverse peoples, I saw the world through my postage-stamp collection, locating countries on a small atlas as I applied the hinge to the back of the stamp to add to my album. I did notice that Great Britain was the only country that was entitled to anonymity on a stamp – everyone should jolly well know who we are even without a label (though the royal head did rather give it away).

Arthur Mee's encyclopedia was about much more than the Empire. It was organised into sections covering many areas of knowledge. Religion was essentially the Church of England, with brief nods to other faiths. The constellations were illustrated by the stars of the night, with spectral great bears or Diana the huntress sketched in to dramatise the heavens. The history of life was in there, too, tucked away between articles on art and architecture, or the origin of foodstuffs, or stories of great people who did great things. It was all distinctly *worthy*. The Wilshin Aunts could smile indulgently while young Richard played the bookworm, knowing that he was in safe hands. Facts absorbed by that growing brain remain obstinately ineradicable. I realised just a few days ago that the reason I know that sago is extracted from the pith of a cycad palm tree is because of Arthur Mee; a fact that has stood me in good stead for many years.

First among the stories of great discoveries was that of Dmitri Mendeleev and his recognition of the periodic table of elements. This was thrilling to an aspiring chemist; it made the ranks of my chemicals more than a list, it slotted them into the grand scheme of things. Elements went into families, and like all families they shared patterns of behaviour. This was the first step towards our understanding of matter at the atomic, and now at the subatomic level. I felt a sense of wonder that Mendeleev was able to predict the discovery of elements to fill the 'holes' that existed in his table. I began to see how common salt, vital to life, sodium chloride (NaCl) itself, could be utterly different from its components: a metal so active that it fizzed on water, and a gas so choking that to manufacture it only once was quite enough for me. I learned of chlorine's wicked younger sister – fluorine, producing a marriage with hydrogen so greedy that it could gobble glass and burn through floors. The encyclopedia told another story of Marie and Pierre

Curie and the isolation of radioactive elements like radium, which I thought of as elements that yearned to leave their own family and join another. Marie Curie labouring with tons of heavy pitchblende to extract the merest whisper of radium showed me that persistence was an important part of science, and Marie Curie's death from exposure to the very thing that made her famous told me what tragedy was long before I had seen *King Lear*.

Art sections of the encyclopedia were rather strong on the late Pre-Raphaelites, particularly Lord Frederic Leighton and Sir Lawrence Alma-Tadema. This was before these painters became unfashionable, and long before they became fashionable all over again in recent years; I suspect this choice may have been Arthur Mee's own from the very first edition. Classical themes were popular and the female form was ubiquitous: Leighton's nymphs often wore curiously wrinkled wraps that adhered to their bodies in interesting ways. One of Alma-Tadema's was boldly nude, frolicking in a bath or a sylvan glade (he was good at water). When I was somewhat older they were my first introduction to eroticism. I suspect I am the only male in existence to have successfully ogled a naiad of Leighton. The Wilshin aunts would have spun in their graves. Arthur Mee would have taken a turn or two in sympathy.

The most complex achievement of the chemi-shed was in attaining the apotheosis of pong. Yet another book was the starting point. *The Guinness Book of Records* came as a Christmas present in the later 1950s. Facts were on offer of a different kind from those in Arthur Mee – all the superlatives: fattest, fastest, oldest, smallest, most expensive.

That the book continues to be published is a measure of how much we like to know such things, despite recent attempts to get a place in the book by people who eat large numbers of pies. I looked, of course, for 'smelliest', and there was the entry: '**Smelliest** substance known to man. Ethyl isocyanide C_3H_5N* is reputed to smell of a combination of decaying meat, rotting cabbage and sewer gas . . .' Irresistible. Surely another gauntlet had been thrown down: to synthesise this Mount Everest of the malodorous. A little more research in the library led me on the route to make this special substance, a trail that involved a few reactions in sequence. The only trouble was that the first step in the whole process demanded potassium cyanide, which had another entry in the *Book of Records* as one of the most *poisonous* substances known to man, one that quickly blocks the body's capacity to assimilate oxygen. It was a favoured component of suicide capsules; in the Berlin bunker it finally extinguished Nazi ambitions. There was no way I could obtain potassium cyanide.

This was to reckon without that old blue encyclopedia of knowledge of all bad things. I was excited and surprised to find a recipe there for making cyanide from available ingredients. The cyanide radical (CN) is present quite widely in nature in tiny quantities – it imparts the taste and smell to bay leaves, for example, and is present in fingernails. The recipe involved roasting parings from horses' hooves with several other common chemicals in

---

* There are those who favour the mercaptans for the title, but they don't have the advantage of being so volatile.

a metal tube, and extracting the result with water. My sister had become horse mad, and owned a pony. After the farrier had been to fit new horseshoes abundant shavings of hoof lay on the stable floor, and my most recherché ingredient was to hand. The Bunsen burner worked on the ingredients until the iron tube fairly glowed. I let it cool before adding water and filtering the result. This innocuous-looking, nearly colourless fluid should be deadly cyanide, but there was no way of telling just by squinting at it. The tricky next step concerned changing the putative cyanide into its twin isocyanide. Although they are both made of nitrogen and carbon, isocyanide (the so-called isomer of cyanide) has the symmetry of the molecule reversed. From the chemi-shed shelves came silver nitrate, easily obtained from a photographer's suppliers. A chemical reaction 'switched' silver cyanide into silver isocyanide, and I was now ready for the last transformation. I am coy about revealing the details but ethyl alcohol (thank you, Mr Ehrbar) was involved, and a final distillation brought the experiment to an end. As a precaution I put a wooden clothes peg on my nose when a meagre stream of drops dribbled from the tip of my Liebig condenser into a very small bottle. I shoved in a cork hard after it was finished. Perhaps a third of the bottle was filled with liquid. When I removed the peg from my nose I did not have to wonder whether my long journey from hoof shavings to ethyl isocyanide had been successful: just the smear around the edge of the cork produced an intolerable stench. It was the summation of all things horrible: rotting cabbage and flesh was the least of it. It was also the smell of success.

I should have stopped there, but I was too pleased with myself. The particular talent of ethyl isocyanide is its volatility. If its receptacle was uncorked for a few seconds the smell could fill, and then empty a room. I took the little bottle, concealed in a briefcase, for a walk around Ealing to try it out. The ABC Tearoom was one target, but I lacked the courage to hang around long enough to see the effects. The Central Line at Ealing Broadway station was more the mark. Here at its terminus the train sat at the station with its doors open for some time, while passengers sauntered on to take their seats and await departure. I walked in one door and out another uncorking at the right moment, then sat on a bench to observe the effect. Muffled cries of surprise and disgust were followed by mass evacuation of the carriage. All from a second or two of exposure to the Prince of Pongs. It was very naughty. I made an exit from the station feeling an impossible combination of thrill and guilt. I was screwing up the courage to take my phial of fetid fabrication into Ealing Grammar School for Boys. Our form classroom had big, old-fashioned radiators and high ceilings. About thirty wooden desks accommodated the boys and their books. It took only two drops of isocyanide on a radiator to convince the French master to evacuate the room. We pushed and shoved to get into the corridor. '*Sauve qui peut!*' said the boy who was top of the class. Schoolboys are always on the lookout for a diversion and here was a big one, so the most was made of dramatic cries of '*Quelle pong, Monsieur!*', '*Ou est le mouchoir de ma tante?*' and 'That was one of Smithers' worst!' I could have been caught red-handed, but nobody thought that the culprit might be in the class. Fortunately, the chemistry laboratory was on the floor above the classroom, and Monsieur Marney assumed that the disgraceful miasma that had driven us from our lesson must have originated from above. It somehow had seeped down into our own atmosphere. He stormed off to have a word with the senior chemistry master. But I never again risked trying out my triumph of synthetic chemistry on school premises.

*   *   *

Ealing Grammar School for Boys was an early twentieth-century building on one side of The Green, which lay at the centre of the oldest part of our suburb. It was next door to Walpole Park, where my father collected *Daphnia* for sale at Tooke's. On the other side lay Ealing Studios, a large site that produced a string of comedy classics after the war. *The Ladykillers* of 1955 was my favourite film. It not only brought fame to Alec Guinness, but also established the careers of his fellow character actors, who would go on to appear in a dozen films over the following decade. Long-running television shows were produced in Ealing and one of our minor schoolboy amusements was to name the actors who scuttled across The Green from the studio to the pub. The first *Doctor Who*, William Hartnell, was one of them, as tetchy in real life as he was on screen. It was not exactly Hollywood, but Ealing remains a working studio, and claims to be the longest running in the world. This is more than can be said for Ealing Grammar School for Boys.

Every day I walked to the school from Ainsdale Road, a

distance of more than a mile. School uniform was compulsory, and shoes had to be polished every morning. We were, after all, the boys who had passed the 11-plus, part of the elite, with a shining future ahead of us. I usually had to buff up my toecaps in turn on the back of my socks to polish off the scuffs I had acquired along the way. It was a pleasant enough stroll except on cold winter days, but in 1962 a great smog turned it into an adventure, even an ordeal. This was the revenge imposed on London for two centuries of burning coal. I recognised the sharp tang of sulphur that made the yellow smog so unbearable. The news described it almost cheerily as a 'pea souper', easily countered by a resuscitation of the Dunkirk spirit, but asthmatics were dying all over the city in a dreadful choking blanket that hovered motionless until the wind freshened. Visibility was so reduced I fumbled and groped my way back from school in the dark, progressing from street light to street light, fuzzy in their thick misty envelopes. I stumbled gratefully back through the front gate of 40 Ainsdale Road like an Arctic explorer escaping a blizzard. This was chemistry all gone wrong. Legislation did finally abolish all the coal fires that stoked the smog. The air became much cleaner. Many years later I hosted a Russian academic who had been brought up on Soviet propaganda and the novels of Charles Dickens. 'Where', she demanded, 'is your fog?' I explained that London had become a cleaner city altogether thanks to enlightened democracy. She looked sceptical. 'In that case, what has happened to your boy chimney sweeps?'

Ealing Grammar School was a bit like Arthur Mee made

pedagogical. Self-improvement through knowledge was the very essence: to the deserving the rewards. If it was unashamedly elitist – Miss Jean Brodie might have recognised some aspects – it was also not snobbish. Although the majority of the pupils were from the middle classes there were a number of working-class children who were not made to feel outsiders. They, too, had passed the test. School uniform did genuinely rub out differences in background, and there were none of the expensive gadgets asserting subtle differences in status that were to become part of a competitive race much later. I felt quite comfortable there, and regret for the City of London School soon faded. There was homework – compulsory – that had to be handed in on time. I resented it at first, but now I see it as a preparation for life, for the duty to respect deadlines regardless of how you might feel. The lesson applies whether you work for the *New York Review of Books* or Allied Cogs & Screws, or set out to write 500 words a day. We grammar school boys did not often refer to the institution that educated those who had failed to pass the 11-plus; this was the local secondary modern school at Drayton Manor. We were, as we were told quite often, 'the cream' floating atop a much larger volume of ordinary milk. This was long before milk became homogenised and the metaphor worthless.

The presiding spirit – even designer – of Ealing Grammar School for Boys was the headmaster, A. Sainsbury-Hicks. He always referred to it as 'my school' as if it were a ship, and he was the captain. More disgruntled pupils might have replaced this with prison – and governor. He was a small man, and my father would have described him as 'cocky'. He marched around his school with a proprietary swagger. With his crinkled hair, neat moustache and pinstripe suit he looked as if he might have been in the army, an impression he did nothing to discourage. Most of us boys were scared of him. To those who suggested his bark was worse than his bite I countered that his bark was so bad that a bite was hardly necessary. Unruly boys sent out of the classroom to stand in the corridor ran the risk of the headmaster taking a perambulation around his school, and the sinner becoming the subject of interrogation. He was not inclined to let worms off the hook. How such miscreants longed for the bell that marked the end of the lesson, when worms could crawl away to gym or German! Sainsbury-Hicks was also sole wielder of the cane as the ultimate deterrent – this was a thin whippy affair applied to an outstretched hand. Corporal punishment was not dispensed with Dickensian abandon, but it was always there somewhere in the back of the mind. Chastised and chastened victims who had committed a serious infraction would arrive back in class all red-faced with hints of dabbed-at tears upon the cheeks. Any hubbub in the classroom would momentarily subside.

A. Sainsbury-Hicks clearly modelled his fiefdom on the template of the British public (i.e. private) school. Teachers were addressed as 'sir' and pupils were addressed by their surnames. 'Fortey' attracted some feeble jokes: 'Didn't I teach your older brother Fortissimo?' was one of them. There was an elite within the elite. Prefects were recruited from the ranks of the sixth-formers, and like a pallid imitation of the Etonian 'Pop' they had a special room in which to play games and amuse themselves. Prefects also had the right to discipline younger boys for crimes like fighting in the playground and being remiss with the school cap. In a day school it was hardly possible to maintain an iron grip, but an attempt was made to extend school rule as far as the front door. The cap was a kind of portable symbol of authority, and had to be worn at all times. It was a crime to take off the cap even in the safety of the 65 bus home. Prefects were out there checking up. When the rock 'n' roll era began caps were perched precariously behind a quiff, so that they forever teetered in danger of falling off and being spotted by a spy. These were dangerous times.

As in many aspiring schools the Sainsbury-Hicks measure of success was rather simple. The number of boys that went on to Oxford and Cambridge was a metric that could not be gainsaid, and the more places the better. In the late 1950s and early 1960s Ealing could vie with Manchester Grammar School as the most successful state school sending boys to the leading universities. I arrived as rather an average student, but once I got the idea of work I began to rise through the class like a lazy trout towards a fly. Each year there had to be somebody who was top of the class by virtue of the sum of all his examination marks, and then the list went all the way down to the bottom, one by one. After a year or two I began to feel I belonged in the top three or four. However, when I received a summons from A. Sainsbury-Hicks I was still apprehensive and fearful. What could I have done? His study lay to one side of a corridor leading to the staff room, from which billowed smoke and chatter. I tapped nervously on the door marked *Headmaster*. 'Enter!' he barked. I was surprised to find an unusually benevolent figure, smiling, almost avuncular, rising from his desk. He told me that he had his eye on me to be one of his Oxbridge boys. 'Yes, sir,' I nodded gratefully. He then admonished me that this would require serious dedication in a world full of distraction and temptation. 'Yes, sir,' I agreed, wholeheartedly. I backed out of the room as soon as I could. 'Thank you, sir.' At that moment I became an intellectual (at least for a while). I was probably fourteen years old.

In one respect Ealing Grammar School for Boys was identical to the City of London. Every morning there was assembly for the whole school. Everyone stood to attention, youngest at the front, until A. Sainsbury-Hicks and the masters paraded on to the platform to gaze solemnly at their charges. The headmaster eyed the state of the toecaps of the junior boys as he marched past, like a drill sergeant inspecting the turnout of his squad. If a toecap was defective in the polish department the unfortunate boy was sent out. Sporting achievements were listed with approval. If something bad had happened, like a theft, the headmaster's face empurpled with fury, and he promised that the offender 'would go, and wouldn't come back': expulsion to Drayton Manor would blight his whole future. Everybody held their breath, hoping that the culprit might be revealed by a stifled sob. I am sure that had I been discovered as the perpetrator of the mighty stink I would have been turfed out of the school, regardless of any potential. The ritual part of assembly was thoroughly Church of England, with a reading from the King James Bible, often by a prefect, followed by a hymn, which we always hoped would be 'Jerusalem' but was more often 'Lead kindly light amid th'encircling gloom'.

I loved the language of the King James Bible, and many of its grander phrases are still lodged in my memory, but my sister and I had become sceptical about the existence of the Almighty well before going on to high school. While we were still very young we were packed off to Sunday school at St Peter's Church, a big Victorian Gothic edifice in the older part of Ealing. Nice ladies told us Bible stories at first, which we found quite pleasant, as we didn't then know enough about conception to worry about the Christmas narrative. However, when it came to the approach to Easter we began to have serious doubts. We had received little stickers for Septuagesima and Quinquagesima, and followed the story of Our Lord quite carefully. Rising from the dead just seemed against nature. The Holy Spirit seemed unfathomable. We began to suspect that our parents, who never expressed any religious convictions, simply wanted us out of the way for a couple of hours. I am now convinced that the reason was what the newspapers of the time always termed 'intimacy'. With two businesses to run, Sunday was the day for recreation. We children eventually rebelled in harmony. We hid under the family bed and clung on to the webbing as our parents became cross and grabbed hold of our legs. St Peter's or else! When we expressed our doubts to the Sunday-school teachers they looked genuinely distressed, but after a short phase of half-hearted attempts at persuasion, our visits to St Peter's became a thing of the past. I somehow managed to square my lack of belief with singing hymns with gusto, and enjoying the different music of the Jacobean Bible. I came to appreciate the religious inspiration behind J. S. Bach's compositions, and the paintings of the Renaissance. I just could not bring myself to believe in salvation or its agent.

If I *had* believed in the power of prayer I would have used it to get off games. Every week the class had to board a train from Ealing Broadway station – not the Tube, but the main line – and travel on the old Great Western Railway for a few miles to Greenford and the school playing fields. They must have been a remnant of farmland that did not get covered with houses as London spread westwards. We were shepherded across the Great West Road clutching our games kit. For some boys it was the highlight of the week, a chance to score runs and goals, and points off one another. I rather dreaded the whole business. In the winter it was horribly cold, and since I was as thin as a lath I shivered pathetically. The football teacher had no mercy, suggesting pointedly that if I ran after the ball it might well improve my circulation. This resulted in my haring manically towards the nearest boy with the ball, but for some reason by the time I arrived the ball had somehow always moved elsewhere. It was quite discouraging. On one occasion I did find myself in possession of the ball right in front of the goal. It must have reached me by a series of implausible ricochets. A fearsome paralysis gripped my legs, made worse by cries of 'Fortey! Shoot!' from my team (it could have been 'shoot Fortey'). With a colossal effort of will, I forced my leg stiffly backwards and then forwards like a creaky pendulum and the ball skittered off sideways towards the corner. There was a communal groan. We had a form First Eleven and Second Eleven and I failed to be selected for either, so I joined a few other weeds in the corner of the field pretending to run about. The summer was cricket and at least it was warm. My inability to catch a ball was remarkable. I could run towards the slowest lob into the air that would tumble out of the sky asking to be grasped by cupped hands only to have the ball fall to the ground a couple of feet away. If I was batting, the ball came towards me at an impossible speed and I waved the bat vaguely in its direction as it whizzed past. If the ball hit me, it hurt. The only place I was content was fielding at long stop, or maybe it was extra long stop, far out on the pitch where the ball rarely came, and where I could identify wildflowers and take an interest in passing insects. I thought of the old photographs of my father in the First Eleven of everything at Worcester Royal Grammar School. I recently obtained a copy of the *Worcesterian*, the school magazine, for 1919, and read how Fortey's goals assured wins for the home side in match after match. I wondered where my own ineptitude had come from, and was grateful that my father was never seated behind the goal to see my feebleness for himself. He might have been reminded of tangled fishing lines.

The Art Room came to my rescue. I enrolled to get an extra O level in art, and was allowed to spend sports afternoons free of dribbling badly and missing goals. I suspect such evasion of football and cricket would not be allowed under present rules. Mr Bland was a happy exchange for the games teacher, as he spent a lot of time saying how talented we were. My friend Robert Gibbs was, if such a thing were possible, an even worse footballer than I, and now together we sketched compositions from life, usually a few everyday objects and items of fruit. We learned new words like 'chiaroscuro', 'collage' and 'bas relief'. We had to study nineteenth-century English painting for a written examination, and so became familiar with the works of Turner and Constable and the Norfolk water-colourists. Trips to the Tate Gallery were planned at the weekends. 'Modern art' was still controversial, so when we espoused the genius of Francis Bacon or Pablo Picasso we could feel a little like *enfants terribles* ourselves. This artistic *savoir faire* beat knowing how to do a header or bowl a googly by a mile. Mr Bland kept a large book by Salvador Dalí called *My Secret Life* hidden in a cupboard, and when he was away from the room we took it out and marvelled at its lascivious detail and egomaniacal excess. We even became quite competent at drawing. Although he was always encouraging, Mr Bland told us of a previous student who was so talented that he got 100% in the test; his examiner commented that he could find no flaw in his work. He was Allen Jones, who went on to be a leading figure in the 'pop art' movement and an *enfant terrible* in his own right. The Art Room was a haven, a club where we could be intellectuals, where we could believe we had talent just below that of Allen Jones. We liked it so much we both went on with art as an A level subject, and not just to escape the playing fields of Greenford. Robert Gibbs continued all the way to the Courtauld Institute to become a proper scholar and art historian. I felt there was no contradiction between knowing about chemistry and knowing about art. After all, intellectuals should know about absolutely everything.

*   *   *

The chemistry laboratory in Ealing Grammar School for Boys seemed sterile after my own chemi-shed. It was indeed more *truly* sterile as it had sinks and running water to keep everything spotless. There was a series of parallel benches for the boys to work at and all the glassware one could wish for. Chemicals were labelled, and those that were dangerous were kept in a separate cupboard, but there was still a much wider range on open display than would be permitted today. I think we had to don white coats. Our first chemistry master was Mr Thornhill, a kindly and mild-mannered teacher, if not naturally inspiring. He invented useful chemical mantras for beginners to remember. We tested the acidity of solutions using old-fashioned litmus paper (later replaced by phenolphthalein) as an indicator that changed colour according to whether a test solution was acid or alkaline. The paper came in little strips suitable for dipping into a test tube. In acid solutions the paper was red, and in alkaline ones the colour changed to blue. To help the tyro remember which way round it went Mr Thornhill coined the execrable jingle:

'Red to blue – alkaloo; blue to red – a-sed.' I had already made my own indicators in the chemi-shed from beetroot juice, so I knew all about pH and colour changes. Nowadays, no colour change is involved, and a figure is read off a dial. Nonetheless, Mr Thornhill's rhyme is deeply lodged within, or even below my cerebral cortex. When my poor mother was suffering from advanced Alzheimer's disease I was astonished when she suddenly and triumphantly blurted out: 'The square on the hypotenuse is equal to the sum of the squares on the other two sides!' Pythagoras' theorem had embedded itself so profoundly it had outlasted almost everything else. If I ever suffer similar decline Mr Thornhill's ditty may be the last thing I remember, but nobody will know what it means.

I discovered the cruelty of schoolboys in Mr Thornhill's lab. Early experiments involved the manufacture of hydrogen gas H_2 from hydrochloric acid and iron filings, an equation almost as simple as that describing my first experience with magnesium ribbon:

$$2HCl + Fe = FeCl_2 + H_2$$

The colourless hydrogen was collected into a small test tube and its identity could be checked by the simple expedient of bringing a lighted taper to the mouth of the tube, when the hydrogen exploded with a small 'pop' as it combined with the oxygen in the atmosphere. This immediately suggested to one member of the class (not me) that a much bigger bang could be made in a much bigger flask. To that end rather more iron filings and a good deal more acid were put to work filling a large flask with the gas. When the taper was applied there was a considerable 'boom'. The effect on Mr Thornhill was dramatic: his hands started shaking uncontrollably and he fled from the laboratory. Our form master told us solemnly the following morning that Mr Thornhill had suffered from shell shock during the war, and that sudden loud noises could set him back. The guilty party still thought it was a bit of a laugh.

The chemi-shed smoothed my way comfortably through school chemistry exams. If we were asked to identify mystery combinations of substances I could recognise them at once from Mr Ehrbar's secret provisions. I went through a series of formal identification tests just to get full marks in the exams. The nomenclature of reactions became more complicated, but it was still all about transformation, and my early feel for such molecular trading remained with me. Valency and ions and solubility coefficients were added to my vocabulary, titration to my skills. I learned about catalysts, and how fertiliser could be snatched from the nitrogen in the air itself thanks to the Haber process. Chemical balances could now weigh in milligrams. In the cupboards around the chemistry lab a few molecules were modelled with coloured billiard balls standing in for atoms and with white sticks representing chemical bonds. My recollection is that carbon was appropriately black, hydrogen red, and there was some attempt to suggest their differences in atomic weight by the size of the balls. For a long time I really envisaged that molecules were built out of billiard balls of different sizes held together by electrical rods. Outside, in the real world, Crick and Watson were cracking the secrets of DNA (this was never mentioned in school) but the early models of the most important substance for all living organisms were still constructed from balls and sticks. Then the atom itself was unpicked in our classes into a positive-charged nucleus and surrounding negative-charged electrons, and the nucleus again further anatomised into light protons and heavy neutrons: I envisioned them as packed together like some sort of raspberry. The atomic model arranged as a submicroscopic solar system persisted in my conscious-ness until that, too, was displaced by Nils Bohr's altogether fuzzier but mathematically sophisticated constructions. By now, the chemi-shed was left far behind. Somewhere along the way I lost the simple amazement of burning magne-sium ribbon.

I never became a chemist despite my precocious start. I do not believe this was just because more and more complex ideas and equations erased the simple wonder of transformation. Organic chemistry had its particular fascination – the endless permutations of carbon-based molecules were a new source of astonishment. Carbon atoms could hold on to one another to engender a creative profusion that eclipsed all other elements. I visualised the construction of complex organic chemicals as being as much architecture as science – a benzene ring here, an alkyl radical there, welded together like the iron skeleton of some fantastical skyscraper, trailing oxygen and nitrogen as bolts and rivets, hydrogen as bunting. A whole new language was needed to describe these carbon architraves and flying buttresses. I was soon to learn that carbon's older brother on Mendeleev's periodic table – silicon – could form structures in minerals and rocks nearly as elaborate as those of carbon, another world of complexity and taxonomy. Maybe I felt intimidated by the sheer scale of the task of learning all these new molecular designs, a task for which the chemi-shed had become irrelevant. I persisted with formal chemistry qualifications as far as I could at school but never followed on at university.

This rationalisation of my dwindling motivation is probably disingenuous. I think it is much more plausible that the young chemist felt demoted when he could no longer do it all himself. That is why I recall the details of my own experiments at home with much more clarity than the years with pipettes and white coats in the laboratory at Ealing Grammar School for Boys. I left the romance and excitement of chemistry behind in the chemi-shed. The fact is that it would never get better than making the smelliest substance known to man.

4

The Ammonite

My ammonite is about 165 million years old. It had lain undiscovered, concealed deep in the rocks, until I plucked it from its hiding place. A few years ago somebody accidentally dropped it and split it in two, but the halves still fit together to remake the whole specimen. I would have been dismayed if it had shattered, because it is the first fossil I ever discovered, and I have kept it with me throughout my life. At first glance it looks like a discus, about the size of a tea plate. The ammonite is preserved in limestone, which gives it a creamy yellow colour. It is curled into a flat spiral that expands gradually in width and diameter until the end of its growing margin is over an inch across, the outer whorl somewhat embracing the inner ones. I count five turns on the spiral recording the growth of the animal during its lifetime. Ammonites are an extinct kind of mollusc, and their spiral form compares with the elegant shapes of living ramshorn snails. The names of both snail and fossil refer to their resemblance to horns – the snail reminiscent of those carried by a male sheep, while Ammon Zeus was a horned god in ancient Greece (derived from an Egyptian precursor). My own ammonite carries ribs that run crossways over each whorl in a regular pattern: a strong inner rib is joined by another, intercalated rib on the outside of the whorl. It might almost have been a carving made by a sculptor. A groove runs all along its back on the exterior of the outer whorl. The fossil feels cool to the touch even on a warm day.

Our Stanmore Deluxe caravan had been moved to a site near the sea, where it was made available to let for holidaymakers. Ranks of similar caravans were parked at West Bay, near Burton Bradstock, Dorset, on the English south coast. Our summer holiday was spent in our mobile home, just me, my sister and our mother. Father never joined us, as the businesses had to be looked after, although it was strange how every year the mayfly seemed magically to release him from his duties. West Bay was a rather bleak spot, where a small stream breached the high cliffs and a patch of flat land allowed caravans to provide summer accommodation. The beach was little more than a mass of rounded flinty stones, so there was no chance to build sandcastles or ride donkeys. Even seashells were hard to find. The cliffs were ramparts of rusty yellow sandstones that were fretted into more resistant horizontal benches separated by softer, eroding layers; notices warning of falling rocks were set up along the beach. To reinforce this advice large boulders of hard limestone lay at the foot of the cliffs where they had come to rest after tumbling down from high above. They could have been killers. From the sea's edge it was obvious that the sandstone cliffs were overlain by a prominent band of rock close to the top of the cliff that jutted out in places, ready to drop.

A path led over the cliff inland towards the nearby village, running close to the precipitous edge. A premonition of children falling over the drop made Mother particularly nervous, as we darted cheerfully in all directions over the short turf marking the track. Behind the path and away from the sea was a golf course. It may have been the golfers that were responsible for dredging back some of the rock layer that made the dangerous overhang at the top of the cliff. Now piles of broken limestone rock

My first fossil, the Jurassic ammonite *Parkinsonia*.

lay close to the cliff edge. I saw my ammonite just poking out of the rubble, like a gift specially intended for me and me alone. I grabbed it while my mother shouted at me not to get any closer to the edge for God's sake.

I soon learned the Jurassic age of my discovery from leaflets handed out at nearby Abbotsbury, where I went in my role as bird boy a little later to see the famous mute swans and their nests. The capping rock was known as the Inferior Oolite, a limestone formation that overlay the softer, deep ochre-yellow Bridport Sands making up the bulk of the cliffs. Nowadays, this stretch of coastline is just a small section of a World Heritage Site for the Jurassic Coast. Online visitors can take a walk through time along the coast following the succession of rock formations without danger of falling off the cliffs. They can find out about fossils without getting their hands dirty. For me at that time, it was back to Arthur Mee's encyclopedia, where I turned to an illustration of 'Life in the Jurassic Period'.* Ichthyosaurs and plesiosaurs fought one another in limpid seas, while strange fish lurked in the shallows. Pterodactyls soared in the air above an alien landscape decked out with cycad palms. On the sea floor lay the dead shell of my ammonite. I was thrilled. My discovery was a precious relic of lives long vanished, a message addressed to me from the depths of time. I wanted to know more, to give the fossil a name, to capture it into my personal taxonomy. I

---

* If I recall it correctly, this illustration may have been adapted from a famous nineteenth-century drawing by Henry De la Beche, and so was very antiquated indeed.

did not realise that my encyclopedia's view of the English Jurassic sea was already long obsolete. Nor did this really matter: inspiration was the important thing.

On my first solo visit to the Natural History Museum in London I carried the Dorset ammonite safely in a small bag. South Kensington station was a direct journey on the District Line from Ealing Broadway; I had regularly passed it on my way to City of London School. The museum was reached by way of a passage running below street level, which echoed in a satisfying way – it is still there, and buskers now make use of the acoustics to bulk up their sound. Once I left the tunnel the museum stood before me in all its Gothic immensity, enhanced by its curious towers. The entrance was more like that of a cathedral than a house of science, and I climbed its steps with my heart fluttering nervously. A warden wearing a dark blue uniform and a peaked cap directed me rather grumpily towards an antique, polished brown door. On admittance, an Enquiries Officer in a tweed jacket took my fossil with a friendly smile, issued a receipt, and informed me that it would be examined in due course by a scientist, one who specialised in the Ammonoidea. I was astonished to discover that such a person could exist. Were there really so many ammonites that a scientist could spend a lifetime on their study? I was fond of the Rupert Bear annuals as part of my indiscriminate reading, and Rupert had a friend who was a professor – a benign figure with a bald pate surrounded by a fringe of fluffy hair. My involuntary image of the ammonite expert was similar. I imagined him holding my specimen up to the light as if it were a fine wine, nodding his head wisely, before giving an authoritative opinion in a kindly, if abstracted voice.

Two weeks later I went back to the Natural History Museum – it was called the British Museum (Natural History) at the time – and collected my ammonite. With it was a handwritten label that announced 'Parkinsonia parkinsoni, Inferior Oolite'. My fossil had been identified – or 'determined'. It did not take deductive genius to infer that it was named after someone called Parkinson. My old encyclopedia soon revealed who he was, and, indeed, it was the same Parkinson who first diagnosed the disease that carries his name. But he also wrote pioneering and well-illustrated works on palaeontology, first published in 1804, so my lucky find at Burton Bradstock now had connections not only with the remote Jurassic Period, but also with a famous nineteenth-century scientist. It felt like a propitious combination. As for the Inferior Oolite, I discovered that it was a particular kind of limestone made largely from tiny, perfectly spherical grains called 'ooliths' that could be easily seen under a hand lens. It was 'inferior' because it lay underneath a thicker rock formation called the Great Oolite, not because it somehow failed to come up to scratch. Almost immediately, geology became an additional addiction for me, another mass of facts to master, but facts that seemed appropriate to my other enthusiasms. A collection of fossils carried no worrying baggage about harming living organisms; if I didn't collect them, fossils would crumble away to nothing. I would save precious evidence of the past from destruction. Miss Long would have approved.

Most kids go for dinosaurs. I know several small children who will lecture me solemnly about the differences between *Diplodocus* and *Brachiosaurus*, and master the pronunciation of *Pachycephalosaurus* before they know about multiplication and division. Skeletons of dinosaurs are the centrepiece of every major natural history museum: vast sauropods, toothy carnivores, plated stegosaurs, stocky *Triceratops* with its horns in a threatening pose. Moving animatronic dinosaurs have become more and more realistic. My own grandson hid nervously behind my legs when we went to see the snarling *Tyrannosaurus rex* in London; he was not entirely convinced that it was a simulacrum. For the most part dinosaurs inspire in children just the right mixture of terror and security. They are scarily huge but safely extinct. Smaller, gracile dinosaurs are always on display alongside their monster cousins, but they do not produce quite the same frisson of delicious scariness. Every modern display worth its salt tells the visitor that the dinosaurs did *not* actually go extinct, because they live on as birds, and the best collections will have on display a Chinese fossil that shows an intriguing bridge between the two. Fewer museums have got around to covering their carnivorous dinosaur reconstructions with feathers – for which there is some evidence – it somehow does not fit with their macho *schtick*. Dinosaur-shaped jelly babies are the giant reptiles' saddest commercial demotion – from kings of the Mesozoic world to a sugary novelty.

My first exposure to dinosaurs was thanks to Walt Disney. A trip to what my mother always called the West End was exciting; it was our downtown from suburban Ealing, all smartly dressed people, big shops and red buses. Oxford Street was the nexus of the West End, and somewhere along it a cinema called Studio One was perpetually showing *Fantasia*. The movie was made in 1940, and at the time it was the apotheosis of the cartoon. It is a musical tribute to the possibilities of Disney's medium, as animated episodes illustrate and accompany classical music, conducted by the great Leopold Stokowski. Apart from a mawkish moment when the maestro has to shake hands with Mickey Mouse it is all music and marvellous images: abstract shapes accompany Bach, bucolic rustics illustrate Beethoven's pastoral symphony, mushrooms dance to Tchaikovsky. The episode most people know is Paul Dukas' *The Sorcerer's Apprentice*, where Mickey appears again as the unfortunate hero. However, the most dramatic section is reserved for Igor Stravinsky's *Rite of Spring*. It is used to encapsulate the history of life, albeit a chronologically eccentric history in which the first few hundred million years flash by in a minute or so. Trilobites have a walk-on part for a second, rather in the way that Alfred Hitchcock would appear momentarily in his thrillers. Drama is supplied when the dinosaurs march in. Giant herbivores like *Diplodocus* are shown wallowing in freshwater lakes gobbling weed, an idea current at the time. (Nowadays they are portrayed more like elephants, in great terrestrial herds.) The savage *ostinato* of Stravinsky's *Rite* comes later and is matched by a bloody battle between *T. rex* and *Stegosaurus*. At the end, the whole tribe of dinosaurs lumbers towards extinction as the climate changes and the world becomes a lurid desert. Only bones remain. I was transfixed.

The cartoon had nothing of the veracity of *Jurassic Park* and its numerous successors, but it was relentlessly gripping. My older self might ruminate on the mutability of knowledge – what Disney showed was the science of his time, and much of it has changed. Nothing was known in 1940 of the meteorite impact that is implicated in the demise of the reptilian giants. If it had been known then, it might have supplied an even more dramatic ending for Disney's masterpiece. I understand that reconstructions of dinosaurs will change yet again with new discoveries and new scientific techniques. Maybe what we think we know now will one day line up alongside Arthur Mee and *Fantasia*. That is really not the point. The curiosity stirred up in the young boy endured, even if the narrative has changed in many ways. You cannot calibrate inspiration by a catalogue of facts.

Discovering the ammonite aroused different emotions from those that lit up my day in the cinema on Oxford Street. I could never own a dinosaur, I could only wonder from afar. The Natural History Museum had (and still has) a splendid gallery of marine reptiles – ichthyosaurs and plesiosaurs, several of them collected in the nineteenth century by the pioneering woman scientist, Mary Anning. A number of these specimens were discovered at Lyme Regis, a little further west along the Dorset coast from Burton Bradstock. The most I could expect to find of one of these magnificent beasts would be a vertebra or two. The ammonite was different. It was a whole specimen, and I could add others of its kind to my collection. The commonest fossils are invariably those of smaller marine creatures whose shells were entombed in sediment after they died. They are the pawns on the chessboard of life. Humble they may be, but they constitute a wonderful array of different kinds of animals: clams, snails, sea urchins, brachiopods, belemnites, corals, trilobites. I wondered if there was a specialist for every one of them in the Natural History Museum. I would learn all their names; I would master the past! Fossils would become my next target: I would discover where they were hiding in cliffs and quarries and ditches. Who knows what I might turn up?

There is a common misconception that fossils are uncommon. My first ammonite seemed to me an object of great significance, a rarity as precious as a Roman gold coin. When I began seriously to look for fossils I discovered that there were special localities where they were rather numerous. I soon recognised that a collection could sample a whole ancient habitat, and give a feel for life on sea floors that had long vanished into deep history. I could wade in the warm waters of a Jurassic reef, or I could swim alongside the ammonites. I could even devise my own pictures to replace the old images in *The Children's Encyclopedia*. The charisma of collecting increased in direct proportion to the age of the fossils; the more ancient, the more exotic. I soon learned the geological periods in their proper order, the litany of geological time. Their names seemed romantic and evocative, and the older ones had a special magic. Cambrian, Ordovician, and Silurian had an irresistible allure, an alien strangeness. My new geological map of Britain was easily read as a diagram to predict what fossils I could find in a particular stratum, as the rock formations were plotted out in colour describing bands across the land. Some of the colours on the modern map dated all the way back to the first geological map made by William Smith in 1815, like the green that signified Cretaceous chalk. Green it has remained for two centuries. I still sometimes feel that at some subliminal level chalk must be green, even though I know perfectly well it is as white as . . . chalk.

My father fished pure chalk streams, and on Woodspeen Farm there was an old pit dug down into the white rock beneath, probably to make quicklime to 'sweeten' the soil – part of the process Robinson had outlined in my first chemistry book. It was an obvious place for me to commence fossicking for fossils. Old chalk pits were rather common after the war, a legacy of a time before chemicals were bought in for agricultural purposes. Picking over the pallid lumps in the quarry I soon learned to recognise flints within the chalk; they were part and parcel of the same sediments – hard silica, white outside and black inside, and so inert as to be practically immortal. Sometimes flint nodules resembled a baby's foot or a small head, but they were put there specially to mislead the novice, for these were nothing more than 'pseudofossils', fortuitous shapes designed to fool the unwary. Flints of whatever form survive indefinitely to frustrate gardeners with their unassailable obduracy. Then something different caught my eye, partly sticking out of a lump of chalk, rounded and the size of a large coin, but with a suggestion of pink to it – a fossil! I carried the chalk cobble home to see if I could extract the rest of the fossil from its rocky hiding place. The good thing about chalk is that it is a very soft type of limestone that can even be cut with a sharp knife. I tapped off small chunks that were obviously not part of the fossil, and then carefully pared off layers of the chalk matrix until I was close to the specimen that had lain hidden for so long. When I felt I was in danger of damaging whatever-it-was I stopped carving and started scrubbing. An old toothbrush steeped in water slowly scraped off the chalk, but the bristles were not hard enough to damage any entombed fossil. Gradually, a shape emerged, a heart shape, and as more of the chalk wore away I made out five shallow grooves excavated into the highest part of my find. The opposite side of the fossil was flattened. It was a beautiful sea urchin. I found out later that the scientific name of the heart urchin was *Micraster*, which means 'small star', presumably referring to the five grooves making that rayed pattern. Under a magnifying glass I saw that the whole animal was made up of calcite plates that fitted together as perfectly as a mosaic, except for two round holes that I learned subsequently were the urchin's mouth (on the bottom) and the anus (at the end). The whole urchin would have originally been covered with hairy spines, almost a fur. The glass revealed something else: there were smaller fossils perched upon my fossil: the tube of a small worm, and a branching colony of tiny bryozoans.* These would have settled on the dead urchin to filter out tiny organic particles from the seawater that washed over it. My sea

---

* Bryozoans form branching or mat-like colonies of tiny tubes, each of which holds a tiny filter-feeding animal in life; and yes, there was a specialist in bryozoans behind the scenes at the Natural History Museum.

Fossils of sea urchins were common in the chalk
at the edge of the Berkshire Downs.

urchin housed a whole community! The fossil fitted neatly
into the palm of my hand and I cradled it as if it were
some kind of talisman. I felt that it had crossed 90 million
years especially to be discovered by me, fossil boy.

The collection had begun. I discovered more fossils in
the chalk as I explored other quarries nearby. Two sea
urchins other than *Micraster* found a place in the shoebox
that first housed my finds; one was a neat cone, the other
larger and shaped rather like a round bun. The 'bun' was
cast in flint, but it was no pseudofossil – it looked as if the
inside of the sea urchin had been filled up with flint, as
now there was none of the original shell remaining. I had
picked it up in a field where it must have been brought
up by the plough. Fossil urchins like this were long known

as 'shepherd's crowns', so they must have been spotted by farm hands who knew nothing of their true antiquity. They might well have believed that the 'crowns' were placed in the fields by mischievous sprites. These flinty urchins were much tougher than the original fossils, so they survived erosion and ploughing to pop up from the ground like treasure to puzzle passing shepherds. If the sea urchins were the stars of the collection, there were also simple clams that looked not very different from oysters and scallops, a small ammonite that sported lumps as well as ribs, and rounded brachiopods about the size of boiled sweets. All these discoveries were made of the same shell material – calcium carbonate, chemically the same as chalk itself ($CaCO_3$), the favoured building material for the majority of marine animals. But the quarry was to yield up something utterly different. Out of a lump of chalk a shining, sharp blade protruded – I could have cut myself with it. It was a shark's tooth. The brilliant lustre on the tooth was proof that it was made of bone rather than calcite. It was not difficult to dig out; it was larger than my thumbnail and just the shape you would expect from one of the ranks of teeth that fill the mouth of Jaws. It was going to displace *Micraster* as the favourite chalk discovery. I began to sketch a picture of the chalk sea in my head: sea urchins, brachiopods and clams on the soft sea floor, a shark cruising above in search of fishes. What I could not see was that all the white stuff of the chalk itself comprised uncountable millions of tiny fossils of planktonic, single-celled organisms. It required an electron microscope to explore their secrets.

As the collection grew I needed names for my finds, as

well as a larger cabinet to house it. There was something rather grand about Latin names, and few fossils had common ones. I was not intimidated by classical names because I was taught Latin by Mr Saunders, who drilled his class with such efficiency that nearly everyone got 90%. We learned the translation of Caesar's *Gallic Wars* (Book 3) by heart, so that when it came to our O level examinations we just had to recognise a few prompts to write out the whole passage in perfect English. One phrase that remains with me is that Caesar's navy sails were 'lightly tanned and dressed with alum', for which information I have yet to find a use. Only Smithers misidentified the prompt and wrote out a completely different passage that bore no relation to the Latin original. Few fossil names were as obvious as *Parkinsonia*. Some were real mouthfuls: my heart urchin was probably *Micraster cortestudinarium*, and I confess to a certain braggadocio when I trotted out the name to impress anyone who was interested, and many more who were not. I realised that the complicated species name referred to a heart (*cor*) and a tortoise (*testudo*) so it was not implausible as a description. I had my first inkling that the Latin (I should say scientific) names served as an international language and provided a rational way to deal with the immense diversity of life.

The Geological Museum was next to the Natural History Museum on Exhibition Road in South Kensington. It was much less visited, except by me. The upper floors had systematic collections of British fossils laid out according to their age and locality. Specimens were displayed in old-fashioned cases with glass tops, just ranks of examples with their geological age and scientific name on the label. Southern England would have a case or two with chalk fossils in the Cretaceous section, another couple of cases with the typical fossils from the Inferior Oolite as part of the Jurassic section. This particular case held a much better specimen of the ammonite *Parkinsonia parkinsoni* than my founding fossil. The display was fairly comprehensive, and an enthusiast could pore over the cases almost undisturbed. There was nothing here to get the warders in a tizzy, no 'Do not touch' notices to be disobeyed. I was often the only person in the gallery, and certainly the youngest. I would take out a fossil I had brought with me to match it with one on display. In the chalk section: my pyramidal sea urchin was like *Conulus albogalerus*, my shepherd's crown was a dead ringer for *Echinocorys scutata*. One of my big clams was certainly an *Inoceramus*, but I could not decide which species exactly matched mine. The brachiopods were a more varied group than I had realised: I was learning more precise observation and more refined discrimination. I carried a notebook to record the details so that I would be able to update the labels in my own collection. I was as happy as could be. In the early 1960s the Natural History Museum produced hand-books with good drawings of fossils to supplement my visits* but there was no substitute for looking at the real thing – and besides there was always the chance of finding a species which was not included in the handbook. While my

---

* The Palaeontological Association has since produced a series of excel-lent photographic guides – for example, *Fossils of the chalk*, which would have been perfect for my young days.

contemporaries played Saturday football in Pitshanger Park I was the solitary child in the Geological Museum. My mother must have thought that I was a curious boy.

The display cases have all gone. They began to seem old-fashioned, and they were not as attractive to the average visitor as new exhibits with friendly text offering full explanations, and plenty of special effects. The Story of the Earth appeared in the Geological Museum in the early 1970s and became a great success, because it explained the advances in geological understanding of our home planet since the advent of plate tectonics; it also featured earthquakes and erupting volcanoes. The stratigraphic collection of fossils on display was doomed, and the fine examples of ammonites and all the other invertebrate shells mostly went back into storage in the vaults. Nowadays, there are other ways for curious children to identify their specimens. But if there were such things as ghosts, after the visitors have left for the day I could imagine hearing the footfall of a young wraith walking back and forth seeking the cabinets that once held such a special charm. The former Geological Museum is now just part of the Natural History Museum – the Earth Galleries, or the Red Zone. Schoolchildren flock through it on the way to the dinosaurs.

*   *   *

I was extraordinarily lucky to be able to study geology as a subject while I was at Ealing Grammar School for Boys. The geography master proved to be a geologist manqué.

He was one of those rare teachers that almost every student recalls with gratitude and respect. K. E. Williams was known to most of his students as 'Kew'; for me, he always remained Mr Williams. He was a small Welshman with a long, pointy chin and a ready smile. He never had to raise his voice to control a classroom – it just seemed to happen. Some of his colleagues used tricks to achieve the same end, the deployment of sarcasm being a popular ruse. To a talkative pupil such a master might say, meaningfully: 'I am sure you would wish to share your brilliant insights, Smithers, with the rest of the class.' Smithers blushed, and mumbled in embarrassment as silence fell. Mr Williams just effortlessly interested his audience in the Nile Delta, the crops of the Mediterranean region, or the climate of central Spain. He hardly ever set homework, and almost all his students passed the geography examinations with distinction. His real passion was geology. He had completed a geological master's degree at the University of Aberystwyth. He had published a scientific paper based upon his thesis in the *Geological Magazine* in May 1927: 'The glacial drifts of western Cardiganshire'. The title may not resound mightily, but it was his contribution to scholarship and he was proud of it, shyly producing a 'reprint' of the article for his small geology class to admire. I felt a similar diffidence when I published my first article in what our trade calls a 'learned journal'. Mr Williams had amassed a select teaching collection of fossils that included some of the species I had discovered. This was where I handled my first trilobite.

Day trips into the field in southern England took us over the chalk downs and across the Weald. It was a happy release from the routine of school life; we felt as if we belonged to a small and exclusive club that was allowed to have fun – but we were also learning. This was better than sitting in the physics laboratory fiddling with a Wheatstone bridge! Mr Williams explained how the sequence of strata controlled the landscape. The downs rose up to make the high ground, and the slopes near Box Hill were dark with native yew and box trees, while white gashes marked old quarries that revealed the chalk beneath. I wanted to stop to find another *Micraster* sea urchin. Near Dorking we inspected the Silent Pool; water as pure and clean as in any spa bubbled up from a spring on the line where permeable chalk met the impermeable clay beneath. This Gault clay then made a poorly drained valley that was tracked by the main routes and rivers. It could all be followed logically on the geological map. In the distance the hard ridge of the Lower Greensand rose up at Leith Hill to match the downs in elevation, and we drove to examine it along narrow sunken lanes that had been eroded by centuries of ox carts struggling up steep inclines. Dark green straps of hart's tongue ferns lined the route. From the top of the sandstone scarp the view south over the Weald showed another broad strip of deeply wooded low clay ground with misty hills in the distance where yet more resistant sandstones came into outcrop. Hill and vale followed the bidding of the rocks, and geology provided the key to reading the landscape. Even the older houses we passed reflected the underlying strata: flint with brick corners on the chalk; clay bricks in the poorly drained valleys, timber houses in the wooded Weald. In future I would be wearing geological glasses on my journeys around Britain. Mr Williams had helped to change my vision of my homeland.*

When I was fourteen we had a family holiday (minus Father, of course) to St David's, Pembrokeshire, the smallest cathedral city in Britain, on the south-western tip of Wales. By now, we had graduated from the caravan to staying in a spartan guesthouse. The peninsula on which St David's is perched is rather flat and featureless, and Mr Williams would doubtless have described it as an uplifted peneplain. The city itself was (and is) both ancient and charming, running down the hill towards the old cathedral. Most buildings are rather modest, and almost everything is built from the tough purplish or green sandstone that underlies much of the tip of this small peninsula. It is geologically one of the most ancient areas south of the Highlands of Scotland; the streets of St David's have grown from their rocky foundations and feel at ease with their surroundings. The result is a pleasing little city with nothing that seems out of place. The square-towered cathedral is tucked in a hollow backed by open countryside. It manages to be hard to see until the visitor stumbles right upon it, as if it were crouching discreetly. Ruins of a large and opulent Bishop's Palace add to the impression of an important medieval centre for English church and state when most of Wales was still inaccessible.

The glory of St David's is its coastline, where ancient rocks abut a fierce sea. Inaccessible bays are backed by

---

* I would eventually try to capture what I had discovered in a book called *The Hidden Landscape*, which was published decades later.

mighty vertical cliffs that have been fretted by thousands of years of erosion. Clefts in the rock face make narrow caves into which the waters suck and belch. The strata making up the cliffs have been mangled by the profound earth movements that elevated the Caledonian mountain chain about 400 million years ago: nothing here remains horizontal, and the rock beds are often twisted tortuously into folds, or abruptly terminated by great faults that cut vertically as if to ignore the geology altogether. Different colours were juxtaposed when the earth's crust was sliced into these chunks: Caerfai Bay is backed by bright red shale; Solva displays black slaty rocks; massive, subtly coloured sandstones define promontories. Most implacable of all is St David's Head – a huge mass that welled up as hot magma cooled to make an igneous intrusion that offers endless resistance to the relentless breakers. These rocks have suffered deeply in the past and face an onslaught today; the fine spray thrown skywards as another wave crashes ashore is like the exhalation of some enormous whale. The sea picks at the unyielding rock atom by atom.

On the wall of the guest house an old map illustrated the geology of the St David's area. The rocks had all the glamour of extreme antiquity, for here were the Cambrian and Ordovician strata I had long sought, with their outcrops clearly plotted, and even notes on what to expect in the way of fossils. 'Cambrian trilobites here' was marked on a narrow bay called Porth-y-rhaw; north of St David's Head 'Ordovician graptolites abundant' was printed next to Abereiddy Bay. All I wanted to do was to get to these places to see what I could find. My holiday was about to become my destiny. My sister tells me that it rained a lot of the time, which is not unknown in Wales, but my enthusiasm was more effective in buoying my spirits than any raincoat. All I needed was to be dropped off, and picked up before dark. My mother and sister went off in the car in search of ponies to ride, or to find somewhere to get dry. Porth-y-rhaw hardly qualified as a bay, being more like a diminutive Cornish cove, a narrow cleft in the implacable coastline, with a beach of rounded boulders through which a small stream found its way to the sea. Short turf and gorse covered the higher ground. Dark-coloured hard shales that had been tipped nearly vertical were accessible on the left-hand side of the inlet as I faced out to sea, but the waves broke dangerously close. When I jumped to avoid getting wet I glanced up at the cliff and there was the trilobite – and not just any trilobite, but a huge one, at least a foot long. It was out of reach on the steep rock face, proof if any were needed of the tectonics that had tipped everything upwards. Its surface was wrinkled and cracked, but the trilobite's head, and thorax of many segments could easily be made out even though it was the same colour as the enclosing shale. It had suffered but not been obliterated, a survivor from the Cambrian period, a messenger from 500 million years ago. When I learned it was called *Paradoxides*, it seemed to be an appropriate name: it indeed appeared to be a paradox that such an ancient animal could be so complex, and more so than *Parkinsonia* to my naïve eye. I wanted to find one of these trilobites for myself, but there was no way I could extract a large enough piece of the hard rock. My more modest efforts were not entirely in vain. Once I had learned to break the rock with my hammer along the plane of the ancient sea floor I recovered several tiny trilobites with short thoraces (*Eodiscus punctatus*) just sitting on the surface of the broken rock exactly as they must have died in Cambrian times. Pride of place went to another trilobite, the flanks of whose body seemed to be covered with ramifying veins. I wrapped my finds safely in newspaper so that they could not scratch one another, and this was my first collection of the animals I would spend many years studying. It was a day well spent. When I got back to our lodgings I unwrapped the best specimens trying to imagine the seas in which they lived, taking a dive back through time and space to a world that was waiting to be explored.

Abereiddy Bay was more conventional seaside, wide and gentle compared with Porth-y-rhaw, with a proper beach partly covered with flattened discs of weathered black shale. Beyond it, to the north, a huge 'slate' quarry had carved out a great bowl deep into the cliffs. It is now full of seawater and known as the Blue Lagoon by water-sport enthusiasts and wild swimmers; it was the site of a spectacular landslide in 2018. Derelict buildings made of piled 'slate' lend the place a certain air of romance. Everything about the rocky cliffs is dark, almost brooding. The black rocks are Ordovician in age – the period after the Cambrian – which was a time when marine life radiated into many ecological niches that are still occupied today. When I went to Abereiddy as a youth almost every flat pebble on the beach yielded graptolites when it was split apart. The shales were sometimes a little like piled sheaves of papers that could be parted into single thin layers. The graptolites were not subtle fossils; they were startlingly white against the black background of the rock, and on some surfaces they were so abundant that they smeared into one another. The largest specimens were a couple of inches long at most. The commonest species looked like the tuning fork that was carried by every choirmaster to give a pure tone when it was struck. The inner surface of the 'fork' had a finely serrated edge, and a closer look revealed that each serration was actually the end of an inclined tube, now flattened on to the rock surface. It did not take me very long to add specimens of *Didymograptus murchisoni* to my collection – the only problem this time was that I had to decide which ones to reject.

Graptolites were part of the ancient plankton, floating around the world in the Ordovician seas. They were colonial animals. The little tubes were formerly occupied by tiny creatures (called 'zooids'), that fed on microscopic algae and larvae. There could be dozens of zooids in one graptolite. While trilobites were obviously part of the great phylum Arthropoda – the jointed-legged animals that include living insects, crustaceans and spiders – the place of graptolites in the tree of life was debated for decades. Now they are known to be distant relatives of an obscure living animal phylum called Hemichordata, with a few living examples encrusting pebbles and succeeding very well in being inconspicuous. Their days as graptolites marked the zenith of their evolution, when they would have been both conspicuous and ubiquitous.

The seas thronged with their drifting colonies. Just in Abereiddy Bay there must have been millions of them preserved in the dark shale. I imagined clots of dead graptolites sinking into the depths to a place where nothing else survived, there to be entombed in deep-sea mud forever. Nothing living today is quite like a graptolite. Some years later I spent much time focusing my research on these extinct and curious creatures, trying to understand the apparently endless variation in their colonies. The derivation of the name 'graptolite' comes from Greek for 'writing' – to early observers the fossils looked like symbols written on the rock surface. This was strangely appropriate for the fourteen-year-old boy in Pembrokeshire who tapped rocks with the simple enthusiasm of a besotted naturalist, just rejoicing in what a hammer blow might reveal – to be the first pair of eyes to gaze at white outlines which could have been written messages smuggled down through millions of years. Palaeontology was a road less travelled, an esoteric journey that few others had embarked upon. It had the fascination of the arcane, of being something of a secret world. I felt comfortable there, as I did with Mr Bland and Robert Gibbs in the Art Room, or with Mr Williams on the chalk downs. I felt I was being admitted to another club with a membership I had yet to discover. Maybe the club met behind the old polished doors at the Natural History Museum.

I was repeatedly drawn back to St David's. I came with my school friend Bob Britton during my later teens to find more and better trilobites at Nine Wells, not far from Porth-y-rhaw. Together we explored several other sites in the bays and cliffs, locating the sparse and special layers where fossils could be found. Our collections grew. Even now I feel a twinge of envy for a lovely complete trilobite that Bob hammered out from the Cambrian. When I returned again on a field trip with my Cambridge University class I was an old hand. An old-fashioned and courteous 'gentleman don' – Dr Richard Hey – led the trip; he brought students to south-west Wales every year to give them experience with real rocks. The party trotted around many of my familiar haunts. At Porth-y-rhaw I was incensed to discover that some vandal had mutilated the *Paradoxides* trilobite in the cliff, which was pocked with hammer marks in a bungled attempt to remove it from its rocky perch. It was not now much good for teaching, or anything else. (Nowadays, it is frowned upon to collect anything at all, as the site has been declared a Site of Special Scientific Interest.) When the party trooped around a rocky promontory at Whitesands Bay close to St David's Head, Richard Hey announced that fossils were allegedly to be found there, but none had been discovered in the fifteen years he had been visiting. Insouciantly, I detached myself from the group and went to a particular seam in the cliff face that I had discovered with Bob Britton. I whacked it hard, and extracted a large, net-like graptolite. There was general astonishment. 'Well, I never!' said Richard Hey. In a narrow band of shales at another place, this time near Abereiddy, our leader announced that nobody had found a fossil there since 1896, waving the yellowing pages of a report in the Quarterly Journal of the Geological Society of London as proof. 'Is this what we expected?' I said (I admit a tad *faux*

naïf), handing over a nice example of a 'tuning fork' graptolite plucked from a secret spot. 'Goodness me!' said
Richard Hey. So it continued in other sites, until Dr Hey
had run out of modest expressions of astonishment. Finally
came one site I had *not* visited before, and our leader
remarked that the party might just as well stay at the cliff
top, while I went down and found the fossils. What was
remarkable was that my luck held: I found some odd
brachiopods that had been recorded long ago from the
rocks near the shoreline. I felt that discovering fossils was
my métier – 'Fortey's forte' as Richard Hey put it, smiling
weakly at his own joke. I suspect that my fellow students
had other opinions.

*   *   *

If my sister ever felt that my serial interests tended to take
over both our lives, redress was at hand. Kath had the
young girl's love affair with horses. As time went by the
animals got steadily bigger. She had ponies to ride from a
very young age, and seemed quite reconciled to their
quirks. If one of them was liable to buck her off, she just
climbed back on again. She was quite tiny, so that a good
toss could propel her for some distance, and I remember
her emerging from several hedges cheerfully pulling out
prickles. After one toss near Woodspeen Farm she broke
her arm and the image is ineradicable of this little girl
lying on the ground with her arm bent back the wrong
way at the elbow. One little pony was called Mandy. She
was bought through the magazine *Horse and Hound* which

featured large numbers of small advertisements that lied about the tractable properties of the steeds for sale. Mandy was a sweet-looking roan dedicated to being very disobedient and getting out of places like paddocks. Kath joined the pony club and started to do relay races and pony-back egg-and-spoon races. While I bargained for a piece of chemistry glassware for a birthday present, Kath would plead for items of leather tack like martingales, or for a new snaffle. Mother joined in the horseplay, too, and they were off cantering over the Quantock Hills on holiday while I sat in the car reading the entire collected stories of Sherlock Holmes.

Matters got worse when the jumping started. By now Kath had graduated to a small horse, a dapple grey named Starlight who was something of a 'goer' (14.2 hands is the boundary between pony and horse, one of those facts, like Mr Thornhill's litmus jingle, that is welded into my subconscious). I got used to the sight of my diminutive sister sailing over absurdly high brushwood jumps in the local horse trials, as she tried to knock a second or two off the round. My job was to help with the horsebox and to avoid getting kicked as I followed my sporty sister from show to show. Now, it was my turn to be the assistant. Kath started to acquire rosettes the way I acquired fossils, and I am sure my parents were much more impressed by the former.

Kath's last horse was Curlew, who was a great jumper, and marked her pinnacle as a rider. I did learn to ride myself, eventually, on an old grey called Robert, who may have had a previous life as a carthorse. His back was broad and his fetlocks hairy; his temperament was equable bordering on asleep. I did learn to stay on, turn left and right, and sit at the canter. I enjoyed hacking through the Berkshire Downs following behind Curlew and getting to places I had never reached on foot. I made only one serious mistake. The long track known as the Ridgeway was one of the places we often rode along together, and once I suggested that we swap horses. As soon as I was on Curlew he took off at a tremendous gallop heading for Silbury Hill about twenty miles away. I learned the origin of that expression about getting the bit between the teeth, and there was nothing I could do about it. Kath had no way of propelling Robert fast enough to keep anywhere near. As my sister fell further and further behind she yelled out: 'Bring him round!' So I did. I yanked as hard as I could on one rein, and the well-schooled horse did a surprisingly sharp turn through half a circle. I continued onwards through the air following a well-understood principle of physics. It was fortunate that the Ridgeway is grassy rather than flinty. I was thoroughly winded, and lay on the ground making extraordinary hooting noises while Curlew caught up with a bit of grazing. I never suggested a horse swap again.

Splitting a rock to reveal the remains of an ancient animal is to unearth a secret, something hidden from view and brought into the light. I did not anticipate that Ainsdale Road could spring a similar surprise. I was looking for notepaper in the old desk that held all kinds of family stuff when I noticed a folded document jutting out slightly from a pigeonhole; it had a kind of official look. Curiosity got the better of me. It was a divorce certificate: my father had had a previous marriage. As always, nothing had been said. When I thought about it at all I had assumed that parents were an immutable entity, largely there to serve the needs and wishes of their children. My real business was making fireworks or gathering collections, and the permanence of those who made these activities possible was simply assumed. How could it be otherwise? I needed an explanation. When I confronted my mother with the piece of paper she was quite taken aback. I believe she would have preferred the secret to have remained undiscovered, like a fossil interred in an unreachable part of a cliff face. As far as she was concerned the important part of her history began with her meeting and marrying my father. In the end I learned rather more than I would have wanted. The confession, if that is what it was, introduced me to a strange menu of adult behaviour that I only partly understood.

Constance Sophy Judge was some years older than my father. They must have met when he was working briefly for the United Dairies, doing a job he hated. I was assured I had no half-siblings. They were married in 1931 and were together for five years. When father met my mother there was not much point in pretending that the first marriage would survive. The parting was quite amicable, or so my mother said. Her very conventional parents were shocked that she had taken up with a married man – it simply wasn't the done thing. In those days divorce was quite difficult, and adultery was the only cast-iron escape route. There was a way of getting evidence that would be accepted in court, but leave the good names of both 'respectable' women in the situation untarnished. This required the services of a prostitute, a private detective, and a hotel room in Brighton. My father went with the prostitute to the hotel where they would be registered as Mr and Mrs Smith. The private detective stood outside the hotel all night and reported 'Mr and Mrs Smith' leaving together in the morning, leading the divorce court to an inescapable conclusion. That is all there was to it. Why weren't we told about the first marriage? It was much better forgotten.

However, once unearthed, this story could *not* be forgotten. Maybe it could be obliterated, like the hammer blows that nearly rubbed out the memory of *Paradoxides* in Porth-y-rhaw, but even then the damage would still be visible. And if this story were 'better forgotten' how many more such stories might that apply to? I remembered that book I had discovered high on the bookcase: *The Psychology of Insanity* by Bernard Hart, and the reference to a nervous breakdown. I could now guess when that crisis happened. It must have been after my father left Worcester Royal Grammar School and before he married the first Mrs Fortey. Was this breakdown the reason that he never finished his degree at Oxford? The story we were told was that in those days Oxford had a mandatory Latin qualification, and that my father had one year to complete it – and failed to do so. Another scenario presented itself: the brilliant sports star from a village in Worcestershire was overwhelmed by coming to Oxford, and suffered a temporary collapse. After his recovery, this crisis joined the list of other things about which nothing was said. There is now no 'fossil' to unearth to reveal the truth. What has changed is my view of the perfectionist fly fisherman devoted to his art. Maybe it was that art that kept the other unmentioned and unmentionable things in their place.

When I first visited Brighton I admired its elegant bow-fronted and stuccoed Georgian villas. By the 1960s, many of them were hotels and probably had been hotels for decades. I wondered how many Mr and Mrs Smiths had passed through their polished doors, and imagined the knowing little smile that played on the lips of the concierge as he signalled to the porter to help with the night bag. Every now and then, I suppose, a couple arrived who really *were* Mr and Mrs Smith. Did they have to pretend to be a Mr and Mrs Frobisher to protect their reputation?

5

Fungus

The autumn is my best of times. When the September rains blow in from the west the trees in our woodland are so soaked that raindrops cascade off the leaves on to the forest floor. Dry leaf litter that has lain under the trees for many weeks soon becomes sodden. Mossy banks in the wood that have looked crisp and almost moribund all summer awake within a few days into fresh green cushions that are cool to the touch. Their long wait is over, and now just as the leaves complete their year's work the mosses start to grow and propagate. Rotten logs absorb water into their brown flesh. Smells of rot and growth subtly mesh to infuse the woods with a distinctive fragrance – the combined emanations of freshly dampened earth and all the microscopic life that is grateful for the rain. It is a good smell, one that makes you breathe in deeply. Crane flies bumble through the undergrowth, legs dangling. Myriads of lesser flies emerge from hiding to buzz about to mate or feed. Birds that like to eat insects hop enthusiastically from twig to twig; wood pigeons in small groups scavenge in the litter for morsels to fatten them up for winter. Much of this awakening is invisible, as it takes place below ground when the rainwater finally trickles into cracks and interstices to moisten the soil and subsoil. Uncountable multitudes of woodlice and springtails and mites, and billions of single-celled organisms reinvigorate the soil and provoke roots into action. This hidden world is the concealed vital organ for the health of the wood, its stomach and nervous system. In a couple of weeks a whole kingdom will be made manifest: mushrooms will start to heave up from beneath the moist litter, their caps opening out as they rise, announcing their arrival in colours ranging from gentle tans to startling reds. The fungus season has arrived. It is the time of year when I feel most indisputably alive.

Now is the moment to retrieve the basket from the basement. An old-fashioned basket is just the job, put together from interwoven laths of willow; it has seen years of service. Nowadays, I often lead small parties dedicated to the scientific sampling of British fungi. We call them fungus 'forays'. 'Foraging' is something else. Everything foraged is grist to the saucepan, but a foray will bring home all manner of fungal oddities that might break your teeth or send you mad if you included them in a risotto. The fungus enthusiast will not reject a tiny mushroom growing from an acorn cupule, nor yet discard a black jelly hanging sinisterly from a branch or an orange patch on rotten timber. That is not to say that the pleasures of eating mushrooms are lost on forayers. It is rather that the edible varieties are a fringe benefit. The Kingdom of Fungi is the main event.

When conditions are just right there is nothing more pleasing than walking out on a sunny October day into woodland that you know will be teeming with mushrooms. There are occasions when mushrooms are so abundant it is hardly possible to avoid crushing them underfoot. Nor are any two years identical, especially given the British climate, so every year will produce something that you have never seen before. The only problem with mushrooms is that all the species tend to arrive at once: the better the foray the slower the progress. I know of one forayer who

My old woven basket, used for collecting fungi.

never got beyond the car park. If it is done properly, forayers slowly amble along turning over rotting logs and picking up and examining every small mushroom with a hand lens. It is not aerobic exercise. Oblique shafts of light illuminate the forest floor when the sun gets lower, painting up the glowing caps of red brittlegills, or shining on the glutinous white fruitings of porcelain fungus lined up along old beech logs. Meanwhile, a real specialist might only gather small spots discolouring the leaves of brambles or woodland grasses. When something unusual is discovered a cry goes up and everyone gathers round to admire the find. Names – often scientific names – are bandied about with aplomb. It could be the first time *Rubroboletus satanus* (Satan's bolete) has been seen here for twenty years! Its fat red stem and contrasting white cap pinpoint it at once. Some fungi are collected carefully and put into small jars or compartmentalised boxes to be taken home for examination under a microscope, which is the only way they can be accurately identified. Like the hunt for a superior trilobite in a fossil-rich quarry it is all about the thrill of the chase, except that 'chase' is about the least appropriate word for a rapt traipse through a wood *molto lento*. Nonetheless, there is an element of competitiveness as to who discovers the most beautiful, or recherché, or outstandingly delicious find of the day. Enthusiasm is undiminished as the light begins to fade: a fanatical friend continued to foray by the light of his car headlamps when he was convinced that a really uncommon species had to be out there. It often comes as a surprise to first-timers that there are so many different species of fungi. A good foray in the New Forest easily succeeds in finding more than 150 different species. A largely different set of species – but just as many – would be recovered from a foray in the Highlands of Scotland. Naturalists who devote their lives to birds or orchids, and know every species as well as they know their own family, often pass over the whole Kingdom of Fungi as too prolific, too unpredictable; just too difficult altogether.

I cannot pin down my fungal epiphany. I am sure it was after the bird's egg phase, and I am quite convinced that it ran concurrently with the chemi-shed. This young naturalist and scientist found *everything* worth attention; rather than analyse a part of nature, I sought to learn it all. This was not a conscious ambition – it was rather an instinct that could not be denied. It is as if I were no more than a pair of perceptive, but disembodied eyes observing everything – not unfocused, but voracious. Art and architecture, books and birds, fossils and fungi, were part of a portfolio that knew no limits. Charles Darwin's *Autobiography* reveals that he, too, was restlessly hungry to know about everything. He started on beetles. He became a geologist: on the *Beagle* voyage he made fundamental observations on the geological history of South America, and the origin of oceanic atolls. His visit to the Galapagos Islands – the finches and giant tortoises he discovered there – fuelled the theory of the origin of species by natural selection. He devoted several years to a study of barnacles. Orchids, primroses and worms were grist to his intelligence, and all yielded to his brilliance. My only resemblance to the great man is in the range of my early curiosity; I cannot make similar claims about its depth.

If I cannot exactly remember when I first became engrossed with mushrooms I can remember *why*. They seemed so extraordinarily alien. They appeared so rapidly, and disappeared again with equal dispatch. One day in late summer an old stump in the garden at Ainsdale Road was surrounded by a dozen or so brown mushrooms, with caps shaped like closed umbrellas. A day or two later some of the umbrellas had opened further out, but by the following day many of them had turned into a black mush, and it was not long before all evidence of the mysterious visitors had completely vanished. I wondered how they fed themselves and how they grew so fast. I thought that if they disappeared so quickly they must have achieved what they had set out to do, but I had no idea what that might be. A few weeks later they were back again, if not in such great numbers. My mother said they were nasty toadstools* and not to touch them, but it was too late. I had already picked one, which had fallen to pieces in my hand, revealing that the cap sheltered many thin dark parallel sheets underneath that I would soon learn to call gills (I was used to seeing fish gills, so I could see how this name came about). The toadstool had so little substance, yet I could see where bark had been levered away from the tree by the action of a single toadstool pushing through. I soon found other toadstools to compare with the first:

---

* There is no scientific difference between mushrooms and toadstools. Most people tend to use 'toadstool' to embrace inedible and poisonous mushrooms. In this case all toadstools are mushrooms, but not all mushrooms are toadstools.

small pale ones appeared in the grass in the lawn. When we were by the River Lambourn I discovered a huge, white puffball, larger than a loaf. It looked like a balloon that had somehow been exhaled from the ground among the nettles. It was a fungus – but was it a mushroom? It was so different from the ones around the stump it was hard to see how they could be the same kind of organism. Both mushroom and puffball were imbued with mystery, and all mysteries demand to be investigated.

A book was needed. As always during the 1950s the book was one of the *Observer* series. *The Observer's Book of Common Fungi* by Elsie M. Wakefield was first published in 1954, which does place an approximate date on the awakening of my interest. It would be hard to overstate the importance of these little books to budding naturalists in the post-war years. They were compact, small enough to fit into a schoolboy's pocket. They were generally written in a straightforward way by people who knew their stuff. They were well illustrated for the time, although by modern standards the pictures seem rather small. They crammed a lot of information into 200 or so pages. And the books were inexpensive. I had the butterfly volume, and the ones on freshwater fish, birds, and wildflowers, but for these there were other options. The mushroom volume was the only guide of its kind. There were specialised scientific monographs, to be sure, but these were out of reach to the lay reader. Elsie M. Wakefield was a pioneering woman scientist – a professional mycologist. The names I learned from Wakefield's book have stuck obstinately in my memory and refuse to be displaced by those preferred by later classifications based on modern understanding. I still find myself blurting out '*Hypholoma hydrophilum*' to a bewildered (and much younger) fellow forayer when confronting a dense mass of chocolate brown caps on a decaying log. 'Do you mean *Psathyrella piluliformis*?' comes a rather discombobulated reply. 'The same thing!' say I, confidently. It is just that it has changed its name, but not its identity since the days of the *Observer's Book*, which is what often happens with scientific taxonomy. Somehow, calling this toadstool the 'common stump brittlestem' – a name invented for it recently to sidestep a supposedly popular aversion to scientific names – seems less satisfying than announcing the whole classical mouthful. The common mushrooms that made up most of the entries in the *Observer's Book* are still the ones that turn up regularly on forays, and include the best edibles as well as the most lethal. It provided a good grounding.

I did my own solitary forays thanks to London buses. Red double-decker buses served the city and its suburbs, as they still do today. Beyond that network green buses crossed from the outer suburbs into the countryside. You could get to most places around the perimeter of London by bus: smokers and people who liked to see where they were going went upstairs, and the latter included me. London Transport had a bargain ticket called a Green Rover that allowed you the freedom of the whole network, and the Young Traveller Green Rover was even better value. I must have been about thirteen when I took my basket to Ruislip Woods one autumn with my copy of Wakefield's little book tucked into my pocket. Some of this ancient woodland has since been gobbled up by housing, but then it was an extensive tract of beech, oak and other broadleaved trees, with at least one conifer plantation. Paths led into deep rides, and even from the top of the bus I could see mushrooms dotting the roadside verges. I set off into the wood with my senses tingling, with the same nervous awareness that I felt when seeking subtle signs of hidden birds' nests.

The first large fungus I gathered near a birch tree was one of the most poisonous and also one of the most iconic. The red-capped mushroom 'with white spots' appears on ceramics, Christmas cards, T-shirts and cakes. When a gnome needs a seat this mushroom provides it. When Wiffly-Piffly Bunny goes to visit his friend Mrs Mouse she lives in one of these mushrooms, with little windows opening up in the cap, like attic dormers. I suppose it is an icon of sorts, if only on account of its unmistakable signature. Its common name is fly agaric, and its scientific one *Amanita muscaria*, and for once the latter name has remained inviolate. Finding it was a wonderful moment for the young mycologist. It was difficult to believe that any natural colour could be so brilliant, as if some sprite of the woods had come in the night with a magic paint pot to colour it up. I checked with my little book to learn that the 'spots' can actually be wiped off the scarlet cap – they are removable scabs rather than part of the cap itself. They start as part of a white 'bag' that encases the whole toadstool, and as it grows to scarlet splendour the bag splits into fragments, a few of which remain to decorate the surface. I checked that there was a collar, or ring, on the stem, and that the gills were white. One example went into the basket.

Onwards, down the track to see pale brown, funnel-shaped mushrooms that when picked spilt white milky juice on my fingers. There were only a few mushrooms that had this property, and a search through the *Observer's Book* found the name *Lactarius*, with a milky name to suit. From woody twigs arose little delicate bunches of mushrooms with conical caps, almost too fragile to pick. Their stems were no wider than a knitting needle: I took one home for a more leisurely examination. A log on its side seemed to erupt with stacks of tiny brackets, striped concentrically like decorated awnings. There were fungi everywhere! Under a beech tree, scarlet brittle gills (*Russula*) had caps almost as bright as the fly agaric's, a troop of mushrooms crouched close to the ground like lurid cakes that had popped out of the soil. The one I examined had white gills, and had no ring on its stem, which was about the same size and colour as a stick of blackboard chalk, and snapped as suddenly when it was broken. A very pale yellowish fungus sprouted from the path side, which had a similar stature to the fly agaric, and seemed to show comparable features of the gills and stem. Another *Amanita* – the false death cap. It was supposed to smell like cut potatoes, and yes, that seemed to be the exact analogy. The next page in my book showed an illustration of the true death cap (*Amanita phalloides*), the most lethal toadstool known to man, its cap a sinister green colour. Elsie Wakefield said it was to be avoided at all costs. It must have an entry in *The Guinness Book of Records*. I found mushrooms with chocolate-brown gills and rust-coloured gills, and even one with deep pink gills. They must surely all be different species. My basket was getting full quite quickly, but there seemed no end to the variety. On the ground what looked like a mass of discarded orange peel proved to be yet another kind of fungus, but this one could never be called a toadstool. What name to give it? Then hidden in the beech litter a troop of something entirely black covered a yard or so of the forest floor. The fungi were dark and flaccid funnels a couple of inches high blanketing the ground. They were from another reality, and surely wore the livery of the Underworld, malevolent and unnatural, like an eldrich vegetable from an H. P. Lovecraft story (I had just learned that word 'eldrich'). It was easy to find in the *Observer's Book*, for nothing else resembled it: *Craterellus cornucopioides*. I thought a cornucopia was a horn bursting with good things, and 'horn of plenty' was one of the fungus's common names, but I think I preferred the French *'trompette de mort'* (trumpet of death) as more appropriate to its sombre colours and alien form. I had to take some of these fungi home to scare the family. To my surprise, they were described as 'edible and good' in the little book.

I doubt whether many parents would now allow a pubescent boy to foray alone through dense woods on the outskirts of London. It is the kind of place where madmen dig their shallow graves. It may be a site for clandestine appointments. It would be strangely appropriate if drugs were traded next to the *trompettes de mort*. My parents were not particularly lax. At least, I did get the advice never to talk to strangers. To those who had survived the Blitz I suspect all other dangers shrank in proportion. Let the children run free! I cannot remember being afraid of more than getting lost along the rides. I certainly was *not* afraid of the fungi, but others were. On my way home on the top of the bus an old gentleman wearing a flat cap eyed the basket on my lap. 'Don't touch any of them toadstools, son,' he said in a serious voice, 'or you'll be going to an early grave.' I looked at him with equal seriousness. 'It's all right,' I said, 'this is *Craterellus cornucopioides*.' I pointed at the red cap pointing out of the basket. 'The poisonous one is the fly agaric.' I doubt he believed me.

Once I was back in Ainsdale Road I spread out my finds and tried to identify the mushrooms that I had not had time to name in the woods. It was my first attempt to learn the cast of characters in a play that I had yet fully to understand. Fungi were neither animals nor plants, they were a kingdom all of their own. No wonder they exhibited so many colours and shapes. What I had collected were just the fruit bodies of the fungi, their means of reproduction when they shed their spores, which were far too small to see with the naked eye. I could imagine a brown smear from the gills of a dark mushroom must have been made of thousands of minute spores. I needed Mr Morley-Jones's microscope to see them properly, as they measured just a few thousandths of a millimetre. Spores were like minute seeds that blew in the wind until they landed in exactly the right place to germinate. The business part of the fungus was a mass of white threads that sometimes adhered to the bottom of the stem – this was the spawn (mycelium) that spread through leaves and wood feeding on the substances that plants – *real* plants – had synthesised while they prospered in the sunlight. Fungi were recyclers, cleaners, or occasionally unwelcome guests. As I moved between the illustrations in the *Observer's Book* I began to realise that I had collected some things that were *not* in the book. If they were the right colour the gills were attached in a different way, or the mushrooms bruised red when they were handled, of which there was no mention by the thorough Elsie M. Wakefield, or they gave off an exotic smell. There must be many more mushrooms out there to be identified. I realised I was just at the beginning of a long journey.

My mother left me alone to get on with it, because the television was showing the prime minister, Harold Macmillan, doing something important. Although she regarded herself as unconventional, my mother did like her politicians to be good-looking men with neat moustaches. Her liking for Sir Anthony Eden (Macmillan's predecessor) was based almost entirely on his impeccable manners and his moustache. She preferred Tories, naturally, on account of their better grooming. Following her example, when I was very young I liked the look of Joseph Stalin, who not only had a good moustache but was kind to little children like myself. I could see him beaming at them on the television. I thought he would make a much better 'uncle' than Eddie and Arthur, serving in the shops in Fulham and Willesden. My mother failed to warm to Labour politicians, not, probably, because she differed strongly on principle, but because they were deficient in the manners and moustache department. When Michael Foot (Labour) appeared on the box or on the radio she would describe the politician acidly as a 'whippersnapper'. My sister was also not enamoured with fungi at first, as her ponies and our beloved Shetland sheepdog Berry (successor to Sue the fox terrier) occupied much of her time. Much later Kath became rather skilled at finding ceps, king of the edible mushrooms, and a kind of unofficial competition has been running between us ever since. My father was away with rod and line.

What I loved about identifying mushrooms was the involvement of nearly all my senses: sight, smell, touch and taste. The first principles had been established by pioneering nineteenth-century scientists, before the microscope had become indispensable, so the young mycologist was treading in the same footsteps as his naturalist ancestors. Sight was involved, obviously, as I had to check not just the colour of the cap, but more importantly that of the gills, and stem. The gills were attached to the stem in various ways, sometimes running down it, or curving upwards, maybe not even reaching as far as the stem itself. Some mushrooms had a 'ring' on the stem – like the fly agaric – others did not. To find the colour of the spores meant taking a spore print – placing a cut cap gills down on a sheet of paper and covering it with a glass overnight to see what spore deposit had been thrown down by the morning. White was common, but I soon discovered various shades of brown, and pink, or purplish- to jet-black. Some mushrooms changed colour when the flesh was damaged. I soon learned to recognise the blusher (*Amanita rubescens*), a common relative of the fly agaric that flushed pink when its stem was rubbed. Other mushrooms bruised yellow or even black. Touch: some mushrooms felt silky smooth 'like a kid glove' (*Clitopilus prunulus*), others were subtly rough like shagreen. Taste was important: some fungi had no taste at all while others were distinctive. A taste resembling what E. M. Wakefield called 'new meal' was characteristic of some varieties, like the white St George's mushroom, which separated it out from a number of similar-looking species by its overwhelming smell and taste of newly baked bread. Some of the brittlegills (*Russula*) were peppery to taste, or hot, or simply unpleasant. The rule was to nibble a little bit on the tip of the tongue and spit it out when the taste cut in. The latex that dripped from the milkcaps (*Lactarius*) when their gills were broken could be mild, or practically blow your head off. I used to place a drop on my little finger and test it on the tongue. Even this was sometimes an ordeal. I soon learned that superficially similar mushrooms could have dissimilar tastes. The mycological world was complex and I was finding my way through it by experiment, mushroom by mushroom. As for smell, it was the most difficult of all. Odour can only be described by reference to something else: smells like frying bacon, honey, or blue cheese. Only a few are beyond question. The sulphur knight (*Tricholoma sulphureum*) smells of coal tar – it is *exactly* like freshly laid tarmacadam. Everyone knows aniseed, and a few mushrooms exude a strong odour of ouzo. One of them is also helpfully green in colour (*Clitocybe odora*) – a gentle blue-green, and not the evil green of the death cap. The curious fragrance of freshly cut new potatoes may be subtler, but most people recognise it when it is pointed out, as I did in Ruislip Woods with the false death cap. The scent of radish is common enough in the fungal world, but some people think it is rather more like that of cucumber. Some *Inocybe* species are reputed to smell of sperm. I soon discovered two particularly charming examples. One was the aroma given off by a rather small whitish mushroom that grows in short grass (*Hygrocybe russocoriacea*). Elsie Wakefield said it smelled of Russian leather. What could that be like? I could scarcely go round a leather shop sniffing at articles on sale until I found the right one. Only when I finally, and independently identified the mushroom did I know what Russian leather actually smelled like. The mushroom helped me identify the smell rather than the other way round. I have never met a Russian wearing leather to confirm my diagnosis. My favourite aroma of all was a mushroom that was described as smelling of damp chicken feathers (*Singerocybe phaeophthalma*). I have never managed to verify this by catching a chicken and dousing it with water.

Another book helped me to the science. The seventh in Collins' *New Naturalist* series dealt with fungi. During the 1950s the series was a real trailblazer. The writers were not merely expert, they were often world leaders in research. The book covers have become classics, a blend of the semi-abstract and the specific that has real style. Nikko Tinbergen was the father of ethology – if one excludes Darwin, who inevitably got there first – and *The Herring Gull* was ninth in the series; fourth was L. Dudley Stamp on *Britain's Structure and Scenery*, which was one ancestor of my own book *The Hidden Landscape*. *Mushrooms and Toadstools* by John Ramsbottom was for a year or two my bible. Its combination of history, learning, anecdote and science had me enthralled. I still have my old copy, the dust cover long decayed, the binding coming loose through decades of use.* The book helped me understand the many different kinds of fungi. Puffballs and regular mushrooms were distant cousins, I learned, and there were wonderful things called earthstars for me to discover that looked like fungi crossed with starfish. That 'orange peel' I collected in Ruislip Woods belonged to another group of fungi altogether that bore their spores in flasks (asci – hence the Ascomycetes) rather than exposed to the air on special cells (basidia, hence Basidiomycetes) as in regular mushrooms. A whole chapter dealt with the myco-logical equivalent of ethyl isocyanide – the stinkhorn, the smelliest *fungus* known to man. The stench of rotting flesh that emanated from its dripping gleba attracted flies that helped disperse its spores; its Latin name (*Phallus impudicus*) was a perfect anatomical description of its shape (at that time I would have described it as *rude*). It seemed to me that fungi could resemble almost anything, so long as it was bizarre. I followed every word of Ramsbottom's grisly account of the effects of poisoning by death cap (*Amanita phalloides*): how the victim would recover for a while after a lethal meal only to be plunged into further agonies as

---

* I should mention that *New Naturalist* 96 is a modern one on fungi by Brian Spooner and Peter Roberts. It brings the science right up to date, if lacking the incidental pleasures of the original volume.

the toxin destroyed his or her internal organs. This was the end envisaged for me by the man on the bus back from Ruislip Woods. Within a couple of years I found a death cap for myself – it was hiding among the beech mast and I did not dare touch it. The heart of Ramsbottom's book was a huge list of fungi associated with different habitats that showed me just how many varieties there were that were not in my *Observer's Book*. Several kinds of grasslands, woodlands and fens each had their own special species. It was a challenge thrown down for my future investigations. While I encountered so much that was new I also built up an image of the writer: learned, but with a light touch, kindly, broadly cultured, and literate. Maybe he was like the professor I had imagined identifying my first ammonite in the Natural History Museum. John Ramsbottom had been its Keeper of Botany at a time when fungi and plants were classified together. They are now separate kingdoms; their DNA tells us so.

As for food, I supplied my family with unfamiliar mushrooms, and they ate them. It could be said that eating is the ultimate test of taxonomy, the proof that you are willing to put your money (or mushroom) where your mouth is. I now know enough about fungi to understand that I was really lucky not to have made any mistakes. There are pitfalls that do not appear in *The Observer's Book of Common Fungi*. I might have unwittingly followed Graham Young the 'teacup poisoner' in using my family as experimental material to observe the effects of alkaloid poisoning. There are some deadly toadstools that are normally rare – and hence missed out of any book on common finds. But occasionally they have a 'good year' and appear in several localities. One of the scarlet webcaps (*Cortinarius*) has caused several severe cases of mushroom poisoning, notably in 2008 the family of Nicholas Evans, writer of *The Horse Whisperer*. The guilty mushroom was unusually abundant in Scotland that year, and was presumably mistaken for the excellent edible chanterelle (*Cantharellus cibarius*), which has a similar bright orange colour. Three of Evans' family had to have kidney dialysis; Evans himself eventually received one of his daughter's kidneys three years later. The mushroom that causes more stomach ache and vomiting (but not death) than any other looks very like the familiar 'shop mushroom'. It belongs to the same genus, *Agaricus*, typically with a white cap with black gills and a ring on the stem. Its close relative the field mushroom, *Agaricus campestris*, is the only mushroom picked by a typical forager. Its poisonous and deceptive relation is called the yellow staining mushroom (*Agaricus xanthodermus*, 'yellow skin' in Greek). Even now I will rush excitedly towards a patch of promising white mushrooms only to find that they are this dangerous lookalike, the disagreeable doppelgänger. It does, of course, have a distinctive smell, which is difficult to characterise other than that it is not 'mushroomy'. It is the chemical phenol, which was present in the writing ink we used before the ballpoint pen was invented, and the inkwell became redundant. The identity of the white fraudster is best revealed if the base of its stem is scratched, where the flesh turns canary yellow in a trice. The same often happens with the edge of the cap. Fortunately, this was one trick I learned early on. My neighbour was not so fortunate: he appeared at my door ashen-faced last year. 'What have I eaten?' he asked, holding out a plastic bag with a few mushrooms lying in the bottom. 'I've been sick all night.' The yellow staining mushroom had claimed another victim.

There are relatively few very poisonous toadstools, but many more that are not worth eating – too small, tough, tasteless or vaguely nasty. The *Observer's Book* identified some of the best and safest edible species and these were the ones that found their way on to the family table. Boletes were the easiest to recognise because in this family normal fungus gills were replaced by masses of narrow tubes, giving the underside of the cap a spongy look. There were very few that had to be avoided, and all of these uncommon. The penny bun (*Boletus edulis*) was the most sought after of its kind, as it is everywhere in Europe, where it is variously known as porcini, cep, *Steinpilz*, or Karl Johan's sopp, all of which are synonyms of 'delicious'. It is chunky, and from a distance the cap does indeed looked like a freshly baked old-fashioned bun, its only drawback being that maggots find it as tasty as we do. When you beat them to it, you are treated to a nutty, almost meaty, gastronomic delicacy. The Russians slice them and dry them, and open their jars of dried porcini to inhale the smell to drive away the blues brought on by their endless winters. Parasol mushrooms (*Macrolepiota*) were bigger and more statuesque than anything else, and could be spotted from a speeding car, growing out of rough sward; I would insist on my parents stopping long enough for me to gather the prize. Even the white giant puffball could be sliced and fried in butter when it was young. I was walking in the beech woods near Woodspeen Farm when I discovered my first pure gold: chanterelles. There was a troop of them in a leaf-filled ditch, and they were shining like jewellery: golden yellow funnel-shaped mushrooms variously lobed and pleated, some growing from mounds of moss like baubles presented on velvet cushions. I gathered them joyfully. They smelled of apricots: the Kingdom of the Fungi can always spring a surprise.

And here is another: truffles are fungi that grow to maturity underground. There are numerous species that have evolved this habit, many of them small, roundish and knobbly; few of them are of culinary interest. However, the Italian white truffle (*Tuber magnatum*) is famous for being the most expensive food in the world, outstripping Beluga caviar weight for weight. The aroma of this truffle is a confection of all the most delicious chemicals that can set the salivary glands to work. The black truffle (*Tuber melanosporum*) grows in the south of France and is hardly less delicious; it was formerly so abundant that special trains loaded with them were sent to Paris before the truffle's incomparable odour could fade. The summer or English truffle (*Tuber aestivum*) is not quite in the same class, but it is good enough to have once employed truffle hunters on some of our great estates. Truffles rely on animals with a sharp sense of smell, like wild pigs, to grub them up and spread their spores in their droppings. They are a challenge to find, and those who make their livelihood from truffles use trained dogs to sniff them out. I saw some sorry little dogs in Corsica at their work – they had to be kept hungry to perform, and I was tempted to slip them a few biscuits while nobody was looking. John Ramsbottom mentioned that there were special flies whose larvae gorged on truffles; they, too, can detect their special aroma. The crowning triumph of my youthful forays was discovering my own summer truffle in Savernake Forest, just east of Marlborough. Fine avenues of beech trees divide this ancient woodland. In a clearing I noticed flies dancing over a particular patch of leaf litter – they might be truffle flies (*Suillia*) showing me the site of buried treasure. I dug down, and about six inches below the surface of the ground I found what I had hoped for. In my hand the truffle was about the size and colour of a grenade, all warty on the outside, not a thing of beauty, but to me the most exciting find of my mushrooming year.

I have stayed loyal to both fungi and fossils. The latter became my livelihood, while the former have remained my hobby. It could well have been the other way round.

My mycological knowledge increased in proportion to the books I had available. After Ramsbottom's *New Naturalist* came the Collins field guide, abstracted by F. Bayard Hora of Reading University from a great and beautiful 1935–40 publication in several volumes by J. E. Lange on the fungi of Denmark. My well-thumbed old copy of this Collins guide testifies to the use it enjoyed for many years – it more than doubled the species I could recognise in the field. It was carried in my jacket pocket, and more than once dropped into a puddle (I carefully dabbed the pages dry as if the pictures were by J. M. W. Turner). Lange had followed a tradition of fungal illustration, using watercolour drawings to illustrate the nuances that separated species. He had a knack for encapsulating just the important features that helped the identifier make a decision. F. B. Hora led forays near my home town in which he sent everyone off into the woods to seek interesting mushrooms. Then he would produce a cacophonous blast on a hunting horn to summon his forayers back to base, so that he could deliver judgement on their finds. In 1981 Roger Phillips produced the famous Pan Original *Mushrooms and Other Fungi of Britain and Europe*. This was the first time photographs successfully competed with a tradition of watercolours going back to the eighteenth century. Even Beatrix Potter took time off from *Peter Rabbit* to add some exquisite examples to that genre. Phillips' book described more than 900 species of fungi, using truly excellent photographs. At the time it seemed to be the last word – but of course there is never a last word.

My own interest fluctuated. I never missed a year without going on forays, but while I was establishing my academic career as a palaeontologist my knowledge of fungi ticked over gently. My interest reawakened with the autumn rains, when the basket was disinterred, and gratifying forays through woods began again. I loved introducing wildlife groups to the charm of toadstools. The arguments about smells came up with every new season. There was always somebody who would challenge whether the aroma of the false death cap (*Amanita citrina*) was *really* like cut potatoes, while others would insist that it could hardly be anything else. Nobody disputed that stinkhorns were stinky. There was often an irritating forayer who would ask no question other than 'Can you eat it?' while I was explaining how the gills attach to the stem, or eloquently explaining the function of spore dispersal. I have been known to silence these individuals by inviting them to taste a few drops from a particularly fiery milk mushroom. When I lived in London my local forays were often around West Norwood cemetery, a huge nineteenth-century graveyard not far from our house. I would progress with my basket and hand lens around the trees dotted among rather splendid tombs, poking about and finding all manner of small and interesting fruit bodies. One week I discovered from the local newspaper that a series of drug busts had taken place the week before – the cemetery had apparently been bristling with heroin and cocaine as well as agarics. I imagined what the pushers would have made of my pretence of being a harmless fungus gatherer. It might have been as dangerous as eating a death cap.

Throughout these middle years I was merely revising my knowledge annually when the mushroom 'season' began, recycling what I had learned from Roger Phillips and his predecessors. I was doing just what a nineteenth-century 'botanising' parson might have done. Thirty years ago I decided that I must learn microscopy, and I started to become a modern mycologist. My hand lens was replaced by oil immersion. I looked at spores close up, and at the cellular structure of the edge of the fungus gills. It changed the process of identification, and opened up a new level of accuracy. I could now explore different kinds of fungi: tiny ones that erupted on horse manure; white patches on the underside of logs; even fungi that grew on other fungi. I learned the deficiencies of all my previous books and bought more. In the process I lost something of the innocent pleasure of that young boy who discovered his first chanterelles, or the simple delight of the slightly older one who dug up a summer truffle.

In recent years I have been deprived of my sense of smell. I cannot help wondering whether it is a punishment from the gods for my youthful misuse of the smelliest substance known to man. I can no longer detect Russian leather. I still lead fungus forays but now I have to appoint an 'official nose' before I start. I might choose someone with a resemblance to Cyrano de Bergerac, or if Cyrano has not turned up I find young women are often superior in the nose department. 'Does it smell of wet chicken feathers?' I will ask. 'Oh, absolutely!' comes the reply. Occasionally, I have an independent-minded official nose who disagrees with everything I say. 'Cut new potatoes,' I will announce, holding up a false death cap. 'Nonsense!' the rebel will respond. 'It smells of umbrellas!' There is nothing to do in these circumstances but to hold a referendum on the smell, and then head with the group onwards into the forest in the hope of finding a stinkhorn.

* * *

Chemistry and fossils integrated seamlessly into my progress at Ealing Grammar School for Boys, but I kept the mushrooms to myself. They were my esoteric secret. This was as far away from team games as could be imagined. I was a team of one, rewarding my own curiosity with my own prizes. I never studied biology at school. I thought I could do it all myself, and that it would kill some of the fun I had if it were a 'subject'. I studied almost everything else. As I went upwards through the school system it became clear that I was an all-rounder, which was not a particularly good thing to be (except for getting marks). One or two of my contemporaries were gifted linguists but no good at mathematics. We had one extraordinarily able musician, and Allen Jones the artist has been mentioned. The paths through life of dedicated linguists, musicians and painters could have been printed in their genes. They don't have the agony of choice. In our school lifetime there was one fork in the road that it was not possible to avoid. After the Ordinary level examinations were completed A levels and two years of the sixth form came next, and the school divided into an arts and a science sixth. I was unhappy about closing off a whole range of my interests from more school-time exploration. By the time I reached my teens writing had already become important to me, and a growing cultural life ran in parallel with my natural-history enthusiasms. I never recognised the arts/science 'two cultures' schism that C. P. Snow had promulgated when I was thirteen years old, but I knew that I would be forced to make a decision, and that the outcome would close off some possibilities for ever. And it was almost true: it took me much of my life to reintegrate myself into the person I was when I was sixteen.

I dithered. I felt pulled both ways. The English master said that he trusted I would be following him into the sixth form. He was quite sure that this was the right way for me to go. I had taken my French O level a year early and afterwards spent a rather superfluous year quietly reading a few French novels in class (while the latecomers swotted) among which *Tartarin de Tarascon* by Alphonse Daudet sticks in my mind because it was supposed to be funny, and its humour passed me by. I attempted to learn colloquial phrases from our old textbook. Some of them were funnier than Daudet, because they were completely arcane: 'He always wore a broad-brimmed, black felt hat' is a phrase I have yet to employ in France or anywhere else. The French for 'I couldn't help laughing' was (I think) *'Je n'ai pas pu m'empêcher de rire'* – but when I did use it at last in France I earned only a mystified shrug. My French was about fifty years out of date:* when asked where I lived I was trained to reply *'Je demeure en Londres'* which was equivalent to saying 'I dwell in yonder great city'. A school trip to France was an extraordinary adventure at that time, with terrible old ships that had been around since the Great War that tossed and groaned as they breasted the waves, and swilled with vomit. Matelots with Gauloises hanging on their lips made rude remarks that bore little resemblance to the French I knew. Once on land, the toilets were holes in the ground designed for undignified squatting. I loved it all. I had silly fantasies about being a charismatic Frenchman, rather like Jean-Paul Belmondo, lounge-lizarding on La Rive Gauche. It would be hard to give up. Years later I took up smoking

---

* My French turned out to be useful, archaic or not. I examined two PhD theses in French many years later, though I had to prepare the questions carefully. *Je n'ai pas pu m'empêcher de rire.*

untipped Gitanes because of the swirling gypsy on the packet. I still dream about the decadence of those cigarettes.

On the other hand, I had Mr Williams and geology. I could follow my favourite activities as part of my schoolwork, with the guidance of the best teacher in the school. Mr Williams suggested in his gentle way that he would be happy to welcome me into his small club. I could keep up my art, whatever happened. Even then, the advice for an aspiring scientist was to study physics, chemistry and mathematics. I had done well enough so far to make this a viable choice. I had also taken mathematics early, and to everyone's surprise I had done well afterwards with an extra O level in something called 'Further Maths' completed in one year. Perhaps this was one of my talents. Maybe this was the better way to go?

I needed arbitration. Word of my dilemma must have reached A. Sainsbury-Hicks. I was summoned again to the corridor leading to the staff room. After all, if I made the wrong choice it might diminish his school's chances of getting that winning number of Oxbridge places. The headmaster strutted out of his private quarters and eyed me, meaningfully. 'Now, Fortey,' he barked, 'I'm going to ask you a direct question and I want a direct answer.' 'Thank you, sir,' I said, trying to sound up to the response required. The gaze became more piercing, an awl penetrating my very soul, a skewer spearing the clumsy chunks of my self-doubt, ready for grilling. 'Are you more interested in people or things?' said A. Sainsbury-Hicks. I knew I had seconds to make up my mind. My life whizzed past in a few images: trout, birds, chemicals, fossils, mushrooms. 'Things, sir,' I replied, without much conviction. 'Science sixth for you!' said A. Sainsbury-Hicks. Thus was my future decided in a couple of minutes: the points had been switched, and the track was laid down for me to follow whether I liked it or not.

*   *   *

In the winter of 2006 I went on another of my personal fungus forays. Every weekend during the colder half of the year I find time to make a short walk in the chalk hills around my home town in Oxfordshire. I poke around looking for small things that might be easily overlooked. If there is a mild period, all kinds of subtle fungi decide to produce their fruit bodies and grow spores. A ditch might be just the place to look, or an overgrown hedge bank, or maybe a moss-covered log. The small boy with the *Observer's Book* and a basket is now white-haired and with a hand lens dangling around his neck, but it is the same explorer as ever was, curiously probing around to see what might turn up. The winter is a time for cryptic species hiding under bark, little flasks of spores hidden under the skin of dead and dying trees. I can detect them by stroking the surface of bark as would a blind man. The tips of the hidden fruit bodies project slightly, and fingertips are more sensitive than anything else to minute irregularities. These species all need a microscope for identification (they are specialised ascomycetes), and I could not have attempted them before I had my own laboratory. My family refers to them as my 'things on sticks', with gentle eye-rolling.

One of my local sites is at the edge of an ancient manor

called Grey's Court, owned by the National Trust; a tiny road called Rocky Lane runs past it. The verges of the road are overgrown and damp – just the place for fungi. I was searching for small species that like to grow on the stout stems of dead burdock, that tall herb whose seeds hook into clothing with such remarkable persistence. I had already found one or two common small fungi when I noticed a conspicuous bright orange patch a few centimetres long, on one side of a particularly large standing stem where it disappeared into the mossy ground. Under my lens I saw at once that the surface of the patch was a mass of small pores, the ends of short tubes that must have been lined with spore-bearing cells. I knew at once it was something unusual. I took the fungus back home carefully and placed part of the sample in an airtight box with some moss to keep it in good condition. The rest of the sample I rested on a glass slide to allow the spores to drop from the tubes, which took an hour or two. The spores made a white spore print, but there was nothing unusual in that. Under the microscope they were colourless, elliptical in shape and only four-thousandths of a millimetre long, and about half as wide. I squashed a sample of the orange fruit body, which proved to be rather soft, and composed of strongly coloured threads (hyphae). I owned a large book about these kinds of fungi, which are known as poroids. The book was supposed to summarise the current state of knowledge of several hundred species, with determinations aided by keys to sort out contending identifications. I could not get my find to fit exactly with any fungus in the book. I needed to dry the samples – over very gentle heat – to preserve them from decay. If this is done properly all the microscopic features are well preserved for future study. I was in correspondence with a helpful mycologist at the Royal Botanic Gardens in Kew whose expertise outstripped mine, so off the dried sample went for his scrutiny. He, too, confessed himself baffled. There was, he said, one final recourse. The specimen should be forwarded to the authority on these kinds of fungi, Professor Leif Ryvarden of Oslo University. Breath was bated. Judgement was awaited. At last, I heard from Oslo that the orange patch was, as the jargon has it, a species new to science. That explained why it could not be keyed out in the big book! It seemed extraordinary that it was possible to find a relatively conspicuous new species a mile away from my home in the domestic countryside of the south-east of England: another surprise from the Kingdom of the Fungi. A year later the fungus had a name, *Ceriporiopsis herbicola*. Its second, species name refers to the fact that it was found on a herbaceous stem rather than on wood, which is usual for other species of *Ceriporiopsis*. The name became 'official' when it was published with a description and drawings in a scientific journal published in Norway in July 2007 – *Cerioporiopsis herbicola* Fortey and Ryvarden was on record in perpetuity. This is the only fungus to which my name is attached. Over many years, I had coined dozens of scientific names for fossils, but there is something special about naming something that is still alive. Before it gets a name an organism does not properly exist: the name serves to bring it to reality. It becomes an entry in the catalogue of life. It becomes eligible for a biography.

The full mushroom basket.

6

Entr'acte

For most of my school years I rose an hour early so that I could arrive at The Green, Ealing, in time for choir practice. My juvenile voice was as clear as a bell, with that poignant edge typical of boy sopranos. My mother had even tried to get me into the choir school at New College, Oxford, when she thought that I lacked a brain. Fortunately, at that time I also lacked the ability to read music reliably and their choirmaster sent me packing. Ealing Grammar School boys' choir rehearsals became my favourite part of the day. It nurtured my musical aspirations from the soprano days, through the alto period when my voice started breaking, until stabilisation came at the baritone of my maturity. The boy soprano was allowed to sing the intro to the mighty chorus 'Glory to God' in Handel's *Messiah*. I can still do it, in a strangled falsetto, if I am plied with strong spirits. 'And suddenly there was with the angel a multitude of the heav'nly host, praising God and saying . . .' Not a big part, but a significant one. The choir had a regular engagement at the Christmas carol service, when we sang beautiful arrangements of traditional carols by David Willcocks, choirmaster of King's College, Cambridge, and all the parents attended. We looked forward to the arrival of Simper's mother, who was dressed up to the nines, and wore fantastic hats. The music was moving, no question, and I might have had difficulty squaring this with my increasing atheism. I believe I took God out of the equation and put human genius in its place. I did not examine too closely the conundrum that Johann Sebastian Bach wrote sublime music out of profound conviction. I suppose I was just grateful for his depth of belief: don't look at the motor, look at the distance covered.

In charge of the choir was John Railton, who lived for music. He was the only rival to Mr Williams for modest, but absolute dedication to his subject. He never had to bully or cajole; he just carried his choir with him on the back of his enthusiasm. When he was in his thirties he had a cancer which entailed having one of his arms amputated, but he was soon back on the podium conducting with his other arm. He was a modernist, with particular enthusiasm for the music of Benjamin Britten. We sang all of Britten's cantatas during my schooldays. John Railton introduced us to the visceral excitement of Stravinsky and Bartók, and young enthusiasts would listen respectfully to Vaughan Williams in the music room, where they had a machine that played the new, long-playing vinyl discs.

At home, when I was young we had an old machine that played 78 rpm shellac records that came out of brown paper sleeves decorated with an image of a quizzical terrier looking into a gramophone:'His Master's Voice'. Since each side lasted just a few minutes my music was bite-sized. I knew Tchaikovsky's first piano concerto only in morsels, but I sang along to the opening melody while the striding piano chords accompanied me. Short pieces by Delius, or Chopin études played by Arthur Rubinstein fitted more neatly on to one side. Beniamino Gigli sung conveniently concentrated arias by Puccini, and Wagner was represented only by overtures. Some even older records were inherited from the music-hall days, and humorous monologues written by Stanley Holloway were my favourites. I listened over and over to a funny double act called 'Gert & Daisy Make a Christmas Pudding'. Elsie and Doris Waters (G&D) were once great variety stars, now totally forgotten, but their names – real and stage – could have been a catalogue of the Wilshin Aunts. I remember a few 78s by title alone: a number called 'Under the Bazunka Tree'* will remain forever obscure. Maybe the title was more memorable than the tune.

John Railton introduced some taste into my arbitrary musical selection, and made me a lifelong devotee of many twentieth-century composers whose works I first heard or performed in his company. We sang *Belshazzar's Feast*, William Walton's wonderful and frenetic oratorio, and Carl Orff's *Carmina Burana*, and Constant Lambert's *Rio Grande*. My schoolmate John Sivell played the brilliant piano part in the latter; he was destined to be a professional musician.

---

* Thanks to the miracle of the web I discovered a recording of this song (1932), and it truly fails to live up to the promise of its title.

We became involved with performances of contemporary music: Peter Maxwell Davies's *O Magnum Mysterium* was exceptionally difficult, and I will not forget the almost shockingly intense glittering brilliance of the composer's gaze when he came to give us his blessing. Lennox Berkeley's *Mass* was an altogether gentler affair, and the composer was reticent and gracious. The acme of the choir's achievement was performing in the offstage chorus in one of the very early performances of Benjamin Britten's *War Requiem*. I was, of course, an absolute square – I was neither a hep cat, nor did I dig bebop. Rock 'n' roll passed me by like one of those speeding cricket balls that never engaged with my hand or my bat.

My teens were the years of intellectual seriousness. I have never been so serious since. I climbed rather than retreated inwards, reading everything in an unstructured way, making collections of ideas as I might fossils or fungi. My equally cerebral friends and I had pretensions to be young aficionados at the cultural cutting edge. Our music even left Mr Railton behind as we ventured to hear serialist composers, or the continental *enfants terribles*. I made discoveries. The operas of Leoš Janáček had recently been performed under the championship of Charles Mackerras in the 1960s, and I was an early convert. The BBC had a series of Thursday Invitation Concerts, and they were free – although you had to apply for tickets. They were broadcast on the intellectual's own radio channel, the Third Programme, and the presiding spirit was the very embodiment of the avant-garde, Hans Keller. He looked a little like Albert Einstein, only cleverer. His cranium bulged with brains. *Private Eye* regularly mocked his Germanic earnestness. The concerts were held at the Maida Vale studios in London, and there was Hans Keller himself shaking hands with the people that mattered. A few boys in their mid-teens must have seemed slightly anomalous but nobody threw us out. We heard John Ogden play Olivier Messiaen's *Oiseaux exotiques*, and six instrumentalists and a contralto perform a version of Pierre Boulez's *Le marteau sans maître*. There may have been some Hans Werner Henze. I believe I was also at a performance of John Cage's infamous *4'33"*, but since that comprises four minutes and thirty-three seconds of total silence, it is hard to say.

John Railton disappeared from my life after I left school, and the singing that had once been so important to me became occasional. I briefly joined an amateur choir in Berkshire, and performed the usual glees and madrigals with pleasure, but it was not the same as being part of the exciting London scene. The larynx began to rust. I have never forgotten Railton's influence, and the musical tastes I developed at the time that I sang in the school choir have stayed with me. I never thought to hear about him again, but in 2006 I was surprised to receive a letter from Devon from a gentleman who wanted support for his proposal to get John Railton an MBE – the award for those who are not part of the great and good, but who deserve recognition for what they give to others. It is an award by acclamation, and so is probably the one that is actually worth something. John Railton had retired to Devon and made another, wonderful choir. It gave me great pleasure to compose that letter. I asked my correspondent whether he would give me a contact for my former choir-master – I felt that I needed to thank him for his benign influence on my life. Railton replied to say that he could not remember me very well, but he had a recollection of a jolly, red-haired boy (I was both gloomy and dark). This proves that the importance of the student to the teacher is not equivalent to that of the teacher to the student, but I was not very disappointed. After all, his music came first. I learned in 2012 that the MBE had been awarded to John Railton for his inspiring example.

As the teenage intellectual developed, communication with parents almost ceased. My room became a haven, and the Third Programme on the 'wireless' a constant companion. The radio may have been my best friend during puberty. I certainly did not cause any trouble for Mother or Father; I was more like a trainee monk. The BBC was apparently laying things on especially for me, and it was an extraordinary time for culture. I heard what was possibly the first broadcast of Samuel Beckett's *Waiting for Godot*. I knew it was important, but I could not work out why. I wallowed in an early repeat of Dylan Thomas's *Under Milk Wood*. Some lines lodged in my memory: 'before you let the sun in, mind he wipes his shoes' and the night 'starless and Bible black' are still lurking deeply somewhere inside my skull. I still think of one of my favourite composers as 'Johann Sebastian *mighty* Bach'. I owned a published collection of Thomas's radio pieces (*Quite Early One Morning*) that conjured the poet's Swansea boyhood so well, and I practised trying to write in that wordy, musical way that he had made his own. I listened to accounts of his alcoholic end in New York, and part of me was horrified, and the other part thought that this was the way a poet ought to die. I learned his poem 'Fern Hill' by heart. There was a revival of interest in the anarchic plays of Alfred Jarry, and my ear was glued to the radio to learn of the dreadful deeds and goings-on of *Ubu Roi*. I doubt any youth could have been more eclectic.

The 'theatre of the absurd' was all the rage. I caught Eugene Ionesco's play *Rhinoceros* on the Third Programme, a drama that established the genre. A British playwright in the same vein, N. F. Simpson, was soon enjoying success in London with *A Resounding Tinkle*. My first excursion to the live theatre was to see its successor, *One Way Pendulum*. It was a cacophony of eccentric invention set in a suburban house not so very different from Ainsdale Road, and I could identify with the son, Kirby, who was attempting to teach a battery of stolen 'speak your weight machines'* to sing the Hallelujah chorus from Handel's *Messiah*. N. F. Simpson must surely be the godfather of Monty Python. I was inspired to write a play of my own entitled *The Cow in the Attic*, in which a suburban family sit down to supper while a cow's tail, udders, and so on dangle above their heads from a hole in the ceiling. The dialogue had a wee bit of *Waiting for Godot* about it, I dare say. The English master Mr Sheehan read it and was clearly unaware of its masterly

---

* They are now obsolete. The victim stood on a platform in front of a dial. Insert an old penny and the machine would loudly announce your weight in stones and pounds. A comedian of the time quipped that when his wife tried it the machine said: 'One at a time, please.'

place among its contemporary theatrical *oeuvres*. Sadly, no copy of the work survives.

I read in a voracious but random way. If I discovered writers I liked I barged through almost everything they wrote. Rider Haggard, of course, and how I thrilled to the horrific climax as *She*, the anti-heroine, crumbled into decrepitude within a few ghastly minutes. I had an old set of Charles Dickens in blue covers with the famous illustrations by Phiz. They were gobbled up indiscriminately, despite the tiny print size. Now they have merged into a kind of melange where individual characters are sharply focused but the novels from which they came have become blurry. Wackford Squeers belongs clearly in *Nicholas Nickleby*, and may have unconsciously coloured my descriptions of the head teachers in my own life, but the narrative detail of the novel has escaped me. Feckless Mr Micawber and the unspeakable Uriah Heep belong in *David Copperfield*, and both are still vivid in my head, but the chronology in which they play a part is confused; Mr Micawber is defined further by a wonderful cinema portrayal by W. C. Fields. It is not unlike my attempts to reconstruct this narrative of my own life, where events seem to have more substance than time. I discovered the novels of William Faulkner and worked my way through the whole of Yoknapatawpha County novel by hefty novel. Now, what remains? A febrile picture of the Deep South, like a hangover that refuses to go away, and the execrable Popeye from *Sanctuary* still sharp in my memory, as ineradicable as an unresolved trauma. Agatha Christie was reserved for days when I was unwell: her plots always lasted just as long as my illness.

Once I had been directed towards science, I thought I could make up for the missing 'arts side' by myself. I could become a cultured, well-rounded person through my own efforts, armed with my particular species of retentive seriousness. Thanks to the cunning ruse to escape games into the Art Room, the visual arts remained on my school agenda. However, my wayward and unmethodical approach to reading meant that many writers remained unread. I was obliged to follow neither the canon nor the curriculum, so I did not feel compelled to engage with *The Mill on the Floss*, *Tristram Shandy*, or Trollope. I preferred George Orwell, Scott Fitzgerald and Evelyn Waugh. I read Ray Bradbury and John Wyndham and soon saw the possibilities of science fiction. There was a collection of horror stories compiled by Dorothy L. Sayers in the family bookcase that chilled me more than it should have done. Perhaps that is why I followed Sigmund Freud into dreams and Carl Gustav Jung into archetypes. The public library slaked my appetite for reading, which was as greedy as quicklime for water. I still have great chasms in my appreciation of the standard classics to this day, but, in general, my campaign to be a schoolboy polymath was not unsuccessful: it could be easier to become a literate scientist than to be wedded to the humanities and still retain a broad grasp of the sciences. The foibles of educational systems may have generated the 'two cultures' in the first place, but it is not a gap in human nature. We are all creatures of invention and curiosity, not of circumscribed subjects.

Occasionally, I would slip downstairs at Ainsdale Road and sneak into the living room to watch the television, peeping around the door. Nigel Kneale wrote two gripping weekly science-fiction series centred on the scientist Professor Quatermass. They held me as much as the stories in the Dorothy Sayers collection, and I was paralysed with fear as I watched, but oddly unable to avert my eyes. In *Quatermass and the Pit* a bunch of terrifying Martian horrors were trapped in a spacecraft, and when they were suddenly revealed at the end of an episode in all their horrid arthropodan spindliness it touched some deep memory. It might have been of the scuttling creatures I had seen at the gamekeeper's larder when I was very young.

Many intellectually precocious teenage children write poetry, at least those who are inclined to delve into their psyche. I still have a green-covered school notebook holding some of my early and undisciplined verses. The cover says 'Middlesex County Council' which no longer exists, just as Ealing Grammar School for Boys no longer exists. In the absence of any diaries these pages afford the only direct evidence of my thoughts as I moved towards adulthood. Whatever their shortcomings, I have felt a need to take care of them like nothing else from my past. I showed the same tendencies in my taste for poetry as I did for everything else – I established some favourites and ignored much else of worth. William Blake, W. B. Yeats, Alexander Pope and Jonathan Swift appealed to me particularly* – and for almost incompatible reasons. The Augustan poets attracted me by their wit, clarity and formal

---

* T. S. Eliot was always held up as the example of a modern poet at school, and, like everyone else, we read and reread *The Waste Land*.

cleverness; Blake and Yeats were linked by a kind of visionary irrationality. They made magical lines that simply could not be parsed. I followed Yeats into exploring such hermetic arcana as Rosicrucianism and tried to unscramble some of the mysteries of Blake's rambling prophetic books, supplemented by visits to see his extraordinary water-colours in a special room at the Tate Gallery. If my mind was taken by the rational, critical and formal then another part of me was drawn towards something elusive and irrational. I would surely have agreed with Pope when he wrote: 'Nature and Nature's laws lay hid in night / God said "let Newton be!" and all was light.' Yet paradoxically I could also embrace Blake's now overused line: 'the tigers of wrath are wiser than the horses of instruction'. I had an intuition for some grand passion beyond rationality that could burrow down to the truth of existence.

On one occasion my youthful verses went before a proper poet. Maurice Carpenter lived in Pewsey, not far from our country home. My mother had met his wife through a common interest in horse trials, and my sister befriended his son Robin. Maurice had been taken under the wing of those arbiters of taste, the Sitwells, as one of the 'Forties poets' and he had published several slim volumes. When he proved not to be as stellar as his aris-tocratic patrons might have hoped he was dropped with matching rapidity. He became a schoolmaster to make ends meet and to support his family. Philip Larkin would have approved: he believed that poets should disguise them-selves as teachers or vicars or bank managers, just as he worked as a librarian in Hull University. Maurice's wife would have preferred her husband to abandon writing altogether, but he nursed ambitions to be readmitted to the ranks of the leading poets, at a time when Sylvia Plath had already risen to prominence, and Ted Hughes's best work was to come. I somehow finished up reading him a selection of my attempts at poetry, and he was sufficiently impressed to take them away for further consideration. When he contacted me again, he said that my reading was actually much more persuasive than the poems themselves. It was not quite a put-down, but it was a clear indication that I was no Rimbaud. Nonetheless, I persisted with the business of precise expression that marks out poetry. I believed it was first cousin to the exact observation entailed in the recognition of a fungus or a fossil species. In my old folder, some of the poems are even written on the back of a page with a science essay on the other side. Later, through my university years, I struggled to write something with my own voice that was more than intro-spection. I did feel as strong a bond - probably stronger - with some young Cambridge poets as I did with the scientists. My friend Robert Wells became a true poet and a fine translator of Latin classics. For a while, I dared to think that my own verses were good enough to be called poems, and Robert treated them kindly. If I had followed that road it might have led somewhere interesting, but telling yourself you are a poet is not the same thing as being one.

I have discovered one poem in my old collection that records the moment when I moved from leading a kind of double life to pursuing the scientific route. I imagine it was addressed to Robert Wells, and the fact that it was written on a portable typewriter dates it to my Cambridge days. I taught myself to touch-type on a machine called the Brother Deluxe. This was probably the most useful skill I ever acquired: I have written all my books and papers without looking at the keys. The typescript of the verse is fading after half a century. The excerpts from the poem below could almost be a diary entry of a pivotal moment.

Why should I cease to try
This unheard, unproductive art of poetry?
. . .
Reason rhythms, splendour of rocks,
Visions of lava and atoms, thrills to dumbness
The flickering allsorts of emotions . . .
I will not face the livid nakedness
Of truths that scream
Too deep for physicists.
. . .
Creation must not be a duty but
Inevitable, the lava of the mind outpoured
In images upon the chilling air,
The helpless fossils of past fears
Silicified for ever.

I am still pleased with the last five lines. The rest is better read as therapy – running away from the 'flickering allsorts of emotions' that might otherwise have led God-knows-where. The truth 'too deep for physicists' acknowledges the irrationality, but simultaneous profundity of Yeats and

Blake, but also the general angst of youth. I was surprised to read recently that the arch rationalist Richard Dawkins is also a great admirer of the poetry of Yeats, even though what inspired the poet was a terrible mishmash of the theosophist Madame Blavatsky and Irish mysticism. I was not alone in trying to square an impossible circle.

If I could meet my teenage intellectual apogee now I don't know whether I would admire him or feel sorry for him. I would certainly be impressed by his assurance and boundless curiosity. I suspect he would be rather priggish. His knowledge would be imposing but his recitation of it might soon become tedious. Modern, and superficial assessors might mention him as being 'on the spectrum' but my father would probably have said he 'didn't know his arse from his elbow'; that would have been more accurate. I would regret that his fun was so wholly cerebral. There seemed to be no space for letting go, for girls, for pop music, for hanging around by the swings in parks and trying out cigarettes. Clothes were a distraction, unworthy of the true scholar. Not all of this single-minded focus can be attributed to A. Sainsbury-Hicks and his insistence on trying for the Oxbridge entrance. When hormones kick in, so does sublimation – no doubt the young bookworm had read all about it in Sigmund Freud and Havelock Ellis. Shyness popped up like zits. It proved as obstinate to clear up. What price a spotty, gawky, 'square' young intellectual, given to being somewhat sententious, and revelling in things obscure? The same young man also acquired an armoury of words that served his older version well. The attempt to retain broad intellectual sympathies and not become the specialist's specialist was admirable then, and, even if impossible, would be admirable now. Later in Cambridge I encountered the polymath's polymath, George Steiner, and realised that I was as short of true omniscience in the same proportion as I fell short as a poet.

*   *   *

Weekends were almost always down to the countryside, where I could still be a natural historian when I was not being horsebox boy. Primrose Cottage in Boxford had been exchanged for Forge Cottage, Ham; it was a little further to the west, near Hungerford. The cottage was deeply thatched, very old and rather dark inside. Black oak beams crossed the ceilings and upstairs several of the floors sloped in an exciting fashion. One end of the cottage was weather-boarded – it must have been the site of the original forge. A swarm of bees once settled in it, and for a while honey dribbled down the wall into the drawing room. Ham is a tiny village crouching at the foot of the Berkshire Downs, a tucked-away little place, at the edge of three counties, Berkshire, Wiltshire and Hampshire. The nearest high point of the chalk highlands, where Combe Gibbet stands, is just under 1,000 feet above sea level, which is almost mountainous for southern England. Those condemned to hang from the gibbet in past centuries would have been visible for miles. An ancient track running along the edge of the downs was where my sister's horse Curlew attempted a bid for freedom with me on his back. The trout of the River Lambourn were now further away, but those of the

River Kennet were correspondingly close. Another well-known angler, Bernard Venables, lived in the same village. He had written a popular comic strip in the *Daily Mirror* called 'Mr Crabtree Goes Fishing', and later founded the *Angling Times*. Our telephone number was Inkpen 270, which might have been taken as a portent by any aspiring writer. If you wanted to make a telephone call to Forge Cottage you had to go through an operator, who addressed you as 'caller' in lofty tones. There was a village baker called Mr Lansley, a village squire called Mr Brown, and an old pub that was frequented by a scriptwriter for the radio soap opera *The Archers* who picked up his agricultural tidbits by staying in the pub all day. 'I hear mangelwurzels have been havin' a hard ole toim this year . . .' – that kind of thing. Mr Brown was very old, and when congratulated on both his longevity and sprightliness would reply 'You're young, or you're dead.'

A life divided between Ainsdale Road and Forge Cottage was really quite privileged. My mother used to refer to us being 'comfortably off' by the 1960s. With the social antennae of her generation she would say that we were doubtless middle class, but when I pressed her further she was quite insistent that we were not *lower* middle class, and neither were we *upper* middle class, which only left middle middle class. I really wanted to push her to say whether we were lower middle middle or upper middle middle but by then she detected an element of sarcasm in my questions. The social signifiers that established these niceties were really quite arcane, and in our present materialistic age the system that recognises your place in the *scala naturae* by whether you drive a BMW and what size it is, has the great advantage of simplicity. Naturally, the young intellectual despised such signifiers. The conundrum of whether the WC should be referred to as the toilet (lower middle), loo (middle middle) or lavatory (upper middle) was solved by unfailingly referring to it as 'the jakes' (a Shakespearean term, dontcha know). I was guaranteed to annoy my mother by regularly calling the drawing room (upper middle) the lounge. I simply cannot remember which way napkins, doilies and serviettes went, but I am sure there was a status indicator in there somewhere. My sister and her horses guaranteed that she mostly moved

A sketch of Forge Cottage, Ham, Wiltshire,
made by a school friend in the early sixties.

among the upper-middle-class 'county' and farmers who didn't give a damn – in fact, the social hierarchy at the hunt was more or less related to height above the ground. The upper-class Master (red in all parts – notably face and jacket) was often mounted on the most impressive steed, farmers a little lower and more hobbledehoy, with a declining series of ponies below that, and the 'village' on foot following along behind. My politics nudged progressively leftwards.

The lanes around Ham were deeply sunken into the greensand that lay beneath the chalk. (During the great freeze of 1963 the roads filled with snow and the village was cut off from the outside world for more than a week. Mr Lansley was able to shift all his old stock from the village store, including tough old biscuits and Force breakfast cereal.) In the spring, the scent of ramsons hung heavy in the air, wild garlic carried village-wide in the slightest breeze. At that time the downs were still grazed by sheep, and a mass of daisies, vetch and eggs-and-bacon splashed the smooth hills with yellow patches in the early summer. Fossil sponges could be collected from the rocks in our own garden. A short bicycle ride would take me to pits dug out to accommodate silage, and from their chalk walls ammonites and sea urchins were recovered quite easily. My collections were growing, and each new addition had a number with its locality recorded: I was beginning to learn to be scientifically organised. Then I wrote bucolic poems about beech hangers and the antiquity of flint. Or I cycled to Shalbourne, the neighbouring village, to bring back fresh watercress from the spring waters that were clear as rock crystal – a mineral that now I knew was pure silicon dioxide, with silicon and oxygen conjoined in an impregnable cage. At that time my different worlds were cross-linking like chemical bonds, science and natural history contributing to some kind of poetry. It could not last.

Nor did it. A financial disaster overtook the family. Within the space of a few months 40 Ainsdale Road, our suburban anchor for years and home of the chemi-shed, was sold off to the first buyer.

Money had to be raised – and fast. The reason was not made known to the offspring: as so often, nothing was said. Later, we learned that my father had never filed a tax return. He simply ignored what he did not wish to confront. If official-looking papers arrived they went straight into the bin. All those years of increasing affluence were actually built upon not paying anything to the Inland Revenue. We were indeed 'comfortably off' – comfortably off the official radar. The proceeds from fishing rods and gentles and tropical fish just went into the till and into the pocket. Eventually, the taxman caught up with Tooke's and Harding's. A colossal bill must have been compiled by government investigators, and sent to my father with the obligatory scary threats of imprisonment and disgrace if the sum was not paid up in full within thirty days. That, too, was ignored; it was obviously a jolly good time to go trout fishing. The Inland Revenue is made of stern stuff, and continued to demand arrears for astonishing amounts of money. My mother told us that in the end an officer of the Inland Revenue pursued my father all the way to the riverbank. The government employee stood in his suit among the irises and sedges waving a demand for huge payments, hollering: 'What are you going to *do* about this, Mr Fortey?' while Father continued to cast his pale watery nymph and pretend that the horrible man wasn't there. The stand-off went on for some time: fly flicking delicately through the air on to the stream, ranting official waving his papers next to my studiedly oblivious parent, River Itchen gurgling gently onwards as it had for a thousand years. I am unsure whether the expression 'in denial' was current in the early 1960s, but this remains the definitive example.

My mother had to take control. We discovered then that she was always the tougher and more practical partner in the marriage. If the family home in Ealing had to be sold, so be it. Margaret Zander Winifred Wilshin had inherited a portfolio of shares from her father that had already been used several times as security for the businesses during our years of growing affluence. Her father had been shrewd in his investments, and doubtless had good City advice; his acumen saved our bacon. Needs must, thought Mother. Sale of the house and many of the shares was enough to get the man from the Inland Revenue to go away. If a crisis was averted it was also the end of a kind of innocence: the notion of our father as the strong and silent type was hardly consistent with wilful financial irresponsibility. Maybe his early 'nervous breakdown' was not so inexplicable after all, and the silence about it more understandable. If there were parental rows we never heard them, but ledgers and bank balances became part of mother's fiefdom thereafter. If there were tears shed for severing her connection to Ealing after two generations we never got to see them. The 'queen of the suburbs' was history. The countryside would be our home from now on. Who wants suburbia when there is open downland, and a thickly thatched cottage set in a pretty village? My sister could relocate to a good grammar school in Newbury, Berkshire. And it would be much handier for the horses.

I was another problem to be sorted out. I had advanced too far at Ealing Grammar School for Boys to be moved willy-nilly to a new school. Tooke's came to the rescue. Above the premises at 614 Fulham Road was a gloomy, but spacious apartment. I could live there, sometimes with my father when he was in London, and see out my schooldays above a fishing-tackle shop a few miles east of The Green, Ealing. At the weekends I would join the family in the country, commuting westwards from Paddington station to Hungerford, Berkshire; here I would be picked up to give a hand with the horsebox and muck out the stables. It was one way to ensure that I continued my serious trajectory through the education system, and I accepted it without demur.

The Fulham flat was dowdy and neglected, and the decorations dated from well before the Second World War, such that the linoleum and wallpaper had faded equally to an unenthusiastic brown. Hot water came from a fearsome geyser with a booming gas ring beneath it. When it filled the bath, gasps and wheezes and sudden jolting noises suggested that it was heated by an angry dragon hidden deep within the large circular tank hanging off the wall. My bedroom had a high ceiling and just a wardrobe and nothing cheery about it at all. The living room lay at the front of the house, and commanded a view of Fulham Road, lined on both sides with small shops. Opposite 614 on the other side was a thoroughly weird window with a large neon flying saucer illuminated in spooky green and red occupying much of it. It never seemed to be switched off. A sign announced that this was the headquarters of the Aetherius Society. The mission of this organisation was to foster relations with the alien beings now visiting our planet in their vaguely saucer-like machines (today's UFOs). They seemed to be appearing all over the planet. *Plan 9 from Outer Space* is a 1959 movie in which hordes of saucers invade the earth with evil intent, even if they appeared on screen to be jiggling slightly, as if attached to some all too terrestrial cotton threads. The Aetherius Society logo was very like one of writer-producer-director Ed Wood's cosmic interlopers. Occasional visitors in dark coats would scuttle through the side door of the Society's premises, as if bearing secret tidings. If I was attempting to do my homework in the odd green glow that infused the living room at night, my eyes would tend to stray to the building opposite to detect any hint of what went on behind the neon saucer. I never found out anything of interest.

At this time my favourite listening was the Bela Bartók string quartets, and the dissonant fourth in particular seemed to embody the right degree of malaise, as I sat alone wrestling with my physics and chemistry and pure mathematics, bathed in a sickly light from the Aetherius Society. In that eerie pall my face must have resembled the pallor of the dead poet in Henry Wallis's *The Death of Chatterton*. It was perhaps rather unwise to leave a sensitive youth in such gloomy environs, where Mr Railton's 'jolly, red-headed' boy could not have been further from reality, and I did succumb to existential gloom.

Now I was realising that the 'all-rounder' had his limitations. As I advanced further into mathematics I recognised shortcomings in my capacity to think in the abstract. I was fine with geometry, anything I could visualise, but the more sophisticated algebra became, the more I knew that I was applying formulae by rote, just obeying the rules, and that I had not got under the skin of how mathematics really worked. I learned to perform, not to understand. I began to see that physics was, at root, more mathematics, and had been since the time of Isaac Newton. I had never formally studied biology, my natural métier, because I had assumed that I could find out all about that stuff by myself. I had chemistry and geology (and art) to play to my natural inclinations. Much of my life at 614 Fulham Road, however, was a hard slog through calculus and conductivity.

There were diversions. A. Sainsbury-Hicks considered drama an enhancement to the Oxbridge applicant. He also realised that the convention of having small boys playing female leads had its limitations, so he licensed a production of Sheridan's *School for Scandal* in conjunction with the most superior girls' school in Ealing: Notting Hill School for Girls. The actors might genuinely include both sexes. Lady Sneerwell, for example, was actually female. I was Snake, a small, yet I like to consider vital part with some very good comic lines and not too much to learn. In the play I had to double-cross almost everybody, which allowed for a fair degree of overacting in the pursuit of laughs. On the night of the performance Mrs Simper turned up in a hat resembling a large stack of American pancakes. Desperate parents behind her craned their necks to catch a few moments of the performance by their loved one. My friend Bob Bunker played the duplicitous Sir Joseph Surface with panache, and all the laughs came on time. I believe that Lady Sneerwell fancied me a little. I could have invited her back to 614 Fulham Road, with no parental supervision. Somehow the depressing decor of my bedroom would not have encouraged any kind of bliss, or even moderately satisfactory fumbling. Shyness and embarrassment ruled once more.

In the sixth form I was made a prefect. A special room was set aside for this elite recruited from within the elite. This was where I learned the only game at which I excelled: tiddlywinks. There was an official version of this familiar game, with complicated rules. Standard winks came in three sizes – tiny ones about the size of a 5p piece, and others twice the diameter. They were propelled ('squidged') by a still larger wink ('squidger' of course). A large, felt mat (also standard) was unrolled to start the game, and a small pot with sloping sides was placed at the centre. Each player had two of the big winks and four or five of the small ones. It was not a question of just getting the winks into the pot. Most of the tactics were about landing your wink on top of an opponent's ('squopping') which meant the underdog could not be played until released. Complicated multi-wink piles could result, but if your wink was on top you ruled the roost and paralysed everything below. If you managed to squop all your opponent's winks you had as many goes at the pot as you had free winks – and an extra go if you succeeded in getting one in. I developed a way of potting winks even if they were under the rim of the pot. Some games could become very elaborate before all your winks made it into the pot.* Ah! The fun we had! The bad thing about being a prefect was not only the school cap, but having to wear one with a tassel on it. When I was travelling on the Tube or the bus pretty girls would remark loudly to their friends: 'Look at 'im . . . 'e's got a tassel on 'is 'ead!' It was mortifying, but as a prefect, I could not possibly take off the cap.

This period was dominated by changes in perception: my father was not fully in control of our family circumstances; fishing overruled wisdom; I was not quite the 'all-rounder' I had been branded, but natural history remained my consolation and inspiration, a route to dispelling the disquietude I had felt alone in the Fulham Road; our mother had more strength than I realised, but where there is strength there is often also the possibility for confrontation. I soon discovered another, and unsuspected, talent of my father's. Bernard Venables had set up a new, deliberately literary fishing magazine called *Creel* (1963–6). It was an upmarket journal to balance the successful mass-market *Angling Times*, with its cover that

---

* I went on to play for Cambridge University which means I have an unofficial quarter blue. It would not have been sufficient to land me a job in the City.

invariably featured a bloke displaying an implausibly large carp between outstretched arms. *Creel* was more in the tradition of Isaak Walton and G. E. M. Skues in extolling the finer points of the art of the angler. Venables commissioned Frank Fortey to write articles for early numbers of the magazine. I still have copies dating from the middle of 1963. The pieces show a precise command of language with an occasional poetic turn of phrase, and Father's delight in the skills of successful fishing seems to mirror the intense pleasure I had in discovering a fine trilobite. It seems that both writing about, and analysis of, the natural world may have been something we shared. Even if we did not talk about it, there was a common connection at another level, running deep as a hidden pool in a trout stream.

Difficult news awaited me at Ealing Grammar School for Boys. A. Sainsbury-Hicks decided that I should be head boy, presumably because he believed that I was able simultaneously to manage examinations and extra responsibility. I lacked the courage to refuse. It was a jubilee year for the school and the headmaster had ambitious plans to raise money for a new hall, so I was called upon to make short speeches to the mayor and other bigwigs, and to administer the prefects. I certainly would not administer punishments. I enjoyed reading out the most declamatory parts of the Bible from the podium at assembly ('Vanity of vanities, saith the Preacher, vanity of vanities! All is vanity'). However, now I had to see A. Sainsbury-Hicks on a daily basis, so that he could display his glorious ambitions for his school to a deferential pair of ears. I was not very good at the head-boy job, and my deputy, John Banger, would have done it much better. Worse, it was obvious to me that A. Sainsbury-Hicks recognised this fact quite quickly. He sometimes eyed me balefully, as if he could not quite believe that he had chosen me for the top spot. Mr Williams had been right – when I told him about becoming head boy he had said: 'Oh, what a pity . . .' The only good thing to come out of it was that I learned to avoid administering anything for the rest of my life if I could possibly avoid it. A. Sainsbury-Hicks eventually got his new school hall, to add to his stellar university statistics.

My forceful headmaster's achievements crumbled away quickly. Within a few years grammar schools were abolished in the London area, and all A. Sainsbury-Hicks's ambitions to create the best state school in the capital came to nothing. Nowadays, the buildings he strived for are part of a tertiary college, specialising in media, and trading on their proximity to the famous Ealing Studios. *Doctor Who* won in the end.

*   *   *

Tragic events often start innocuously enough. Just before Christmas in 1963 my father gave me a lift back from London to Ham at the weekend on a bitterly cold night. He always drove too fast, with a cigarette clenched between his teeth. At that time the A4 road ran through the periphery of Newbury through a part of the old town with some high red-brick walls. The car hit a patch of black ice, skidded, and plunged directly into a wall. My father probably died instantly. I was in the passenger seat, and survived with a broken arm and some bad bruising. I woke up in hospital. All clear memories of that night have been erased from my mind, and remain so.

My mother never fully recovered from the shock. Although she was not demonstrative, she was devoted to her wayward fly fisherman. They had built up everything they had together. They had survived the Second World War as a team. She had helped save the family's security from the taxman. For years, she had taken two children on family caravan holidays – to Dorset, Somerset and West Sussex – coping alone so that her husband could have a fishing break of his own. There was no funeral. She must have preferred to witness his cremation all by herself. She told us that she wanted to spare us a painful experience, but maybe she did not want us to see how bereft she was. For the rest of her life she trotted out the mantra 'you've got to be tough'. Although it was often directed at some perceived feebleness on my part, I suspect she was speaking to herself much of the time. She quickly disposed of nearly all of Frank's effects that had anything to do with fishing. I believe the rods and the fishing library went to my father's best friend, John Goddard, who became a renowned fishing writer, and acknowledged my father in several of his books. It was as if giving away the angling gear became a symbol of moving on; or perhaps she did not want too many reminders of all that she had lost. She failed to settle in one place for the rest of her life – after leaving Forge Cottage she moved house six times. Taking on a new property was another way of forgetting her bereavement, of leaving an old life behind before its memories could surprise her in the dark. As for me, I would never discover whether I had more in common with my father than I had thought. He went away just as he was coming into focus. He might even have forgiven me for not being a fisherman.

This was my lowest point. I had my A level examinations approaching, by which so much store had been set. I had to cope with those school leadership duties for which I had little talent. I had nowhere to live in London; it was now impossible to imagine living alone in the dingy apartment at 614 Fulham Road. The worst bugaboos haunted me. Arts versus science seemed of little moment compared with life versus death. I kept going by doggedness alone. A kindly family – the Brownings – offered to take me in until I finished my examinations. Their son Charles was one of my contemporaries; they lived in a modest house in Isleworth to the west of Ealing. They were generous and supportive, and I suspect I was very unrewarding, just carrying on carrying on. I do not remember much of my time with these good people; I suppose I was shutting out grief, and most of the world with it. I do recall one redemptive moment. Mrs Browning entered a local flower show with some roses that she had gathered that morning from their own, very ordinary rose bushes. She won first prize, beating rose specialists who had laboured for months over their blooms. Her delight was unbounded, and her pleasure was balm to the desensitisation that inward sadness breeds. I smiled for the first time in several months.

This chapter has not been about science or natural history. It is a kind of entr'acte. I cannot disentangle the story of the scientist I became from a reconstruction of the youth I once was – a complex person, not altogether sympathetic, too serious, voraciously questioning. If it had not been for several strong shoves from A. Sainsbury-Hicks I might not have tried for a leading university, nor would I have necessarily gone along the way marked 'science' when I met that mandatory fork in the road. It took me three decades to bring those roads together again, and to recombine myself; but the poet never returned. I survived the solitary sadness of Fulham, and the sudden death of my father, so there was resilience in there somewhere. The natural world took me out of myself. Science required concentration rather than brooding. The pleasure of discovery is not like anything else, and it cannot be faked. Through all the difficult times, I continued to add to my knowledge of nature. I quarried for fossils or filled baskets with fungi. I cannot say whether I was driven by compulsion or by evasion, but it was fundamental to my sense of who I was: a curious boy.

*   *   *

An odd quirk of my father was that he never washed his hair. Instead, he used some sort of hair oil to slick back his locks every day. This mysterious unguent may have been a descendant of the Macassar oil used by Victorian gentlemen to make their hair smooth and shiny; I never saw it anywhere else. My father's daily toilet was rather particular. He shaved with a razor – the Rolls 'autostrop'* – that could

---

* The vintage 'autostrop' is currently selling online for about £60.

be resharpened after use in a special box, so that a single blade lasted for months. Whatever honed the blade made a curious clattering noise, which was part of the regular ritual. Only the best badger-hair shaving brushes would suffice, and I often wondered how the hair from the badgers could be collected to make them. I had never heard of badger farms. I liked to stroke the dry brush and imagine the living animal, the hair so soft and pliant. After a thick icing of shaving cream had been lathered with the brush the shave itself had the precision of a professional using a cut-throat razor. A quick rinse followed, and then came the application of the pomade (or whatever it was); it was worked into the hair, which was then brushed back vigorously. Completion of the ritual was marked by the ignition of a Player's Navy Cut. On one occasion my mother insisted that his hair just *had* to be washed. She ignored any excuses. Maybe it had something to do with the nits my sister and I picked up from the scruffy kids at Bagnor. The purging of the hair took place over the washbasin. My father reluctantly held his head down as if he were bowing to an execution. Hot water was swished over the greasy locks and an attempt was made to raise suds from a shampoo. A kind of grey sludge oozed off the hair and coloured the water in the basin black. The slurry gurgled reluctantly away down the waste pipe. The process was repeated, and a similar result recurred: grime coaxed from concealment. The washing went on and on again and again until we children got the giggles, and since giggles are more catching than chicken pox Mother started giggling too. Now everyone was laughing fit to bust, including my father, and for no good reason, but the lack of a reason made the whole thing still funnier. Somehow, more sluicing continued, and eventually the water began to run clear. Now the drying process began, and when it was finished my father's hair had turned all grey and fluffy. It was the funniest thing in the world; that laughter still lives on in some remote corner of the universe, an eternal echo from one of the best moments in a vanished life.

7

Flowers

The most dog-eared book I own is now so battered that its grey covers are coming away from the spine. The dust jacket fell to pieces a long time ago and I had to consult the Internet to remind myself what it originally looked like: a lively blue gentian and a nodding fritillary were featured items on the cover, both set against a pink background. *A Pocket Guide to Wild Flowers* by David McClintock and R. S. R. Fitter was another in the line of transformative books published by Collins for the amateur naturalist – an illustrated compendium of the complete British flora. The date in my copy is 1961 so I was fifteen when I started to use it. I began to tick off the plants I had seen from that time onwards, and I have used the same book ever since as a kind of diary recording additions to my own personal flora. My very first entries in pencil are still visible, if faded, and at the outset I did not write down details of sites, although I can remember many of them: the flower name alone triggers precise recall. Later entries are brief: 'Walberswick 1989' and the like, but they, too, often come with sharp visual recollection. I found herb Paris for the first time decades ago in a beech wood on the side of the Berkshire Downs, and the thrill is indelible of finding this uncommon plant among endless groves of dog's mercury clothing the hillside. Since both plant species are green in all parts, including the flowers, the discovery marked a threshold, a point where I knew I was not easily misled by superficial similarities. The book is also a record of how the environment has changed during

Fifty years of continuous botanical use.
A page from my *Collins Pocket Guide to Wild Flowers*,
with two sightings of grass of Parnassus duly recorded.

my lifetime. I have a tick recording the discovery of corn buttercup on Ham Hill long ago, and this species is now red-listed and 'critically endangered'. I can remember it growing by a cattle trough on the edge of a cornfield – a rather small buttercup with a large prickly fruit. I may have been an innocent witness to the passing of a species.

More illustrated books appeared in the next few years to help the amateur botanist, and the flower paintings were often a considerable improvement over the original Collins book. The Reverend Keble Martin's 1965 *Concise British Flora in Colour* has a special place on many a naturalist's shelves, not least because each colour plate is a joy in its own right. This book was more cumbersome to take on a walk than a pocket guide, but when there was a question of identification a small sprig brought home could almost always be matched on one of Keble Martin's colour plates. He sorted out the 'worts' – pearlwort, crosswort and squinancywort, glasswort and pennywort, strapwort and sandwort, stitchwort and fleawort. There were always troublesome plants, like brambles, that seemed to defy identification, and these have continued to be troublesome even in the molecular era, when gene sequences have replaced close observation of prickles or the counting of leaflets. Challenges were part of the fun. The numerous yellow daisies still cause me to confuse my hawkbits with my hawkweeds, and my hawkweeds with my hawksbeards. I continue to feel a sense of achievement if I can negotiate a good identification for a confusing member of the parsley family, as the flowers of this group of plants are all very similar. In most species tiny white flowers are born in flat bouquets called umbels. I used to describe them as uriahs, on account of 'umble Uriah Heep. I learned to call all parsleys 'umbellifers' as did many naturalists of my antiquity (the meaning is 'umbel carriers'). Now we are told that in modern botany they are correctly termed 'Apiaceae', but for me they will remain eternally umbellifers – but this is just my 'umble opinion.

I still return to the Collins pocket guide as the place where I record every plant that I have seen in Britain. I don't know why: it is a ritual I do not question, rather like the Anglican litany of my schooldays. The last addition I made, in 2018, was smooth rupturewort found on a heath in Suffolk, a modest, tiny, and uncommon herb growing over bare soil. It really deserved its mark in the precious old book. At some deep and irrational level I must believe that as long as I keep ticking off more plants in the Collins book I shall live for ever. I cannot possibly leave this world until they are *all* recorded; it is a metaphysical certainty. My tradition of recording goes back to a time before I used scientific names for animals, plants and fungi, so I remember the vast majority of my flowering plants by their old English name – or I should say names, as many of them have a plethora of local tags, as Geoffrey Grigson and Richard Mabey have exhaustively recorded. The charm of wildflower names is often only a short step away from poetry: pellitory-of-the-wall, wood goldilocks, frosted orache, ploughman's spikenard, venus's looking glass. They sound like ingredients for a magic potion. I attempted to teach my children some of these old names, but met a certain resistance. They felt country walks were in danger of becoming more like memory tests. 'Moschatel,' I would intone holding up a very small greenish herb. 'Can't you see it looks just like a town hall clock?' Nobody seemed to disagree. My son found a way out by always giving the same answer. When I held up red campion and asked him to identify it he would reply 'cragwort'; if I displayed henbit dead nettle that was 'cragwort', too; likewise lesser stitchwort. I almost wish there really *was* a plant called cragwort so that there would have been a theoretical possibility of one of the children arriving at the right answer. But of the many worts, none is craggy.

One of the first plants I ticked off in the book was cuckoo flower – also called lady's smock (*Cardamine pratense*). I allow it to grow in the rougher parts of my garden where its very pale pink flowers emerge in small clusters from among coarse grasses, displaying a surprising delicacy among their rank neighbours. Its common name says that the flowers should arrive at the same time as the cuckoo. The annual migration of this avian parasite from southern climes was once routine: as a boy I used to hear cuckoos even in Ealing W5. In 2019 I have waited in vain for the bird to be reunited with its flower: but there is no cuckoo within hearing distance of my small country town, despite the well-wooded hills nearby. The cuckoo population has declined by 80% since I started recording plants in my Collins guide, and research published in 2018 relates this sad loss particularly to adverse conditions along one of two migration routes back to Africa. There is not enough food to sustain the birds on their long journey, and a familiar inventory of causes has been trotted out, yet again, with climate change and unhelpful farming practices top of the list. The arrival of the cuckoo now largely resides as a memory in the name of its charming flower. I dread having to explain what a cuckoo once was to my grand-children, as if it were archaeopteryx. Perhaps I should try to forget the old common name. It is much less upsetting to explain a lady's smock.

British wildflowers have a particular place in my story, as they were the basis of my first scientific paper. As the Cambridge entrance examinations began to loom on the horizon, it was suggested at school that I should try to win a Trevelyan Scholarship. A thesis had to be submitted in support of the application for the scholarship, which offered generous support during the undergraduate years. I decided to work on a project to survey the wildflowers of the chalk downland and adjacent areas along the Berkshire Downs, running from Inkpen Beacon to Shalbourne Hill. This was my local patch around Forge Cottage, and whenever I was at home I went out on the hunt. Intense concentration on the search was balm, or at least distraction, from the events that had shaken my life. I was out in all weathers, sometimes hunched against a strong wind, my eyes scanning the short turf for plants of particular interest. 'Sharp sight' is a curious phrase, but it does convey that the eye spears the sought object, pins it down. In our physics classes the rays that impinged on the retina were always shown as tracks entering the eye from a distance, coming the other way, but the sharp eye reversed the process – the eye did the catching. I thought of a photograph I had seen in Arthur Mee's encyclopedia of an Amazonian native spearing a fish from a dugout. The eye lances the treasure. The hunter can only think of the hunt, no time for morbid thoughts, or self-doubt.

The downlands were in the process of transition when I carried out my survey. The village of Ham was set below the scarp of the chalk, which rose steeply upwards to the ancient track that followed its crest. This was a landscape cleared and exploited by farmers since the Iron Age. For several hundred years the steep slopes of the downs supported sheep and had not been artificially fertilised. The turf was short, and continually nibbled back. Chalk flowers abounded. At some time shortly before my survey began cattle were introduced to increase profitability, and fertilisers were applied to encourage grass growth. I was surveying on the cusp between age-old practice and modern farming. I was not to know that many of the flowers that I recorded with pleasure were destined to disappear from much of the landscape within a few years. At Ham Hill an old, steep-sided cut climbed up the scarp –maybe it was once a drover's route. Now it made a superb habitat for orchids, those most pernickety of flowering plants. I found ten different species. Some were common enough at the time: early purple, spotted, fragrant, and pyramidal orchids, and the all-green twayblade, looking so unassuming among its more glamorous friends. All of them still survive widely in the right habitats. Several species that grew on the steep slopes would now be listed as star performers in any nature reserve: the burnt orchid, compact with a plum-coloured tip to a flowering spike that shaded to white beneath; frog orchid with strange brown-green flowers only remotely resembling the eponymous amphibian; bee orchid on patches of exposed soil, a miniature exotic with a short spike carrying just a few flowers having three pink sepals framing the 'bee' of the petals – the kind of flower that every tyro botanist wants to admire. The Collins book had a measure of rarity for each species, marked by asterisks – zero to three, with three being the rarest. Most of the orchids were 'one star' at the time the book was printed (only the spotted and pyramidal orchids were commoner). Ham Hill yielded further orchids that were flagged with two asterisks, which meant that they were uncommon even in the early 1960s. The man orchid displayed a tall spike of brownish flowers with what the guidebook described as a 'marionette-like lip' – the flowers really did look like a bunch of tiny figures dangling with arms and legs. This unusual orchid was quite common towards the top of the hill and stood proud of the turf – it couldn't be missed. The slender musk orchid was comparatively inconspicuous, with a short spike of white sweet-smelling flowers, and it grew only on the steepest slopes, tucked away. Both were a joy to discover. Even the little lane that led to the hill had a treasure. Leaning out from the hedgebank were long spikes composed of small white flowers: a 'three star' discovery, the Bath asparagus (*Ornithogalum pyrenaicum*). I assume from the name that it was once used as a vegetable, although it is now so rare that if it turns up it makes the news rather than the pot. My survey continued, with a growing inventory of pretty chalk-loving vetches, milkworts and gentians, and many inconspicuous curiosities. Where the flinty ground had been ploughed cornfield weeds were still common – poppies and corn marigolds and venus's looking glass, brilliant scarlet and gamboge yellow washed among the wheat; but that single corn buttercup may have been the last of its kind. I compiled the details of distribution and habitat of each plant to feed into the thesis: it was simple but engrossing science.

I failed to get the Trevelyan Scholarship. The thesis got me as far as the interview, but there I mumbled and bumbled before a distinguished if intimidating board, and I must have failed to impress, or even condemned myself from my own mouth. This scholarship scheme has now been discontinued, but I really want my thesis back. The written evidence of what I saw on the Berkshire and Wiltshire border is a valuable voucher, a benchmark against which to measure what has happened to the flora since the 1960s. It could well provide the only evidence for an important site – I remember that I even included some colour photographs. I have failed to trace any archive of the work submitted to the Trevelyan Scholarship board; I hate to think of those miles trudged and hours with the identification guides simply lost. The long view of nature depends on knowing the status of wildlife over past decades. Memories are fallible. Claims without documentary support about how things used to be might well be treated with suspicion by the next generation. I am reminded of one of my father's eggs, collected from a red-backed shrike's nest in Worcestershire long ago, and now lodged only in my own memory; a sceptic might say it is no more than a tall story.

It may have been a mistake to go back but a few years ago I returned to Ham Hill, to the slopes and banks where I had made my discoveries. From a distance, nothing seemed to have changed: the downs still lay like a supine and sensuous torso below the wide sky. The lane leading to the high ground had been tidied up. When I lived at Ham one of the last hedgers-and-ditchers was still employed by the local council to keep banks in good condition. His tools were a billhook, a sickle and a mattock. Mechanical flails have long since replaced him; I now saw no trace of the rare Bath asparagus. The old track was still there running up the hillside, but it was choked with ash trees, and brambles covered the steep banks that had once been dappled with wildflowers. All the orchids had gone, even the common ones.* At the crest of the hill cereal crops grew almost to the edge of the ancient track. Deeply green fields of barley supported neither poppies nor corn marigolds, nothing was flowering to disturb the programmed growth of cereals. Doubtless yields per acre had trebled over the decades thanks to sprays and fertilisers. I felt bereaved. I recalled that other death: two deaths conflated on a single hilltop commanding one of the best views in the south. My vision north across the green belly of Middle England was blurred with tears.

I have known scientists who regard enthusiasm for the identification of organisms as a kind of stamp collecting. This is not intended to be flattering. They ask: what is the need to know all those damned names? The *real* business is with

---

* I have since heard that the Wiltshire Wildlife Trust is beginning a restoration of Ham Hill and all except the rare orchids can be found again.

sequencing the genome, identifying chemical pathways in organelles, crunching vast sets of data in supercomputers, and other research at the cutting edge. Nineteenth-century vicars did the naming stuff. I have wondered whether some of these critics might regard the extinction of species as rather a good thing, since it would reduce the complexity of natural systems available for analysis. The issue is more than the well-rehearsed division between 'whole organism biologists' and 'scientific reductionists'. I have been on walks with dedicated professional botanists who cannot identify the commonest wildflowers; identification has never been part of their culture. It would be harder for them to experience the empathy with the natural world that I have described earlier in this book. Perhaps they have never felt the harmony that comes with a throng of different flowers buzzing with dozens of insects, a sense of countless natural livings earned in countless ways. Life is polyglottal, symphonic, inventive, and inevitably diverse. Complexity and richness are the hallmarks of life itself. If I could make one generalisation from studying the long history of life on earth it would be that evolution has generated richness. Repeatedly. After the sudden mass extinctions that punctuated life's leisurely trajectory richness reasserted itself every time. After the terrestrial dinosaurs were removed from earth, mammals exploded into a thousand ecological niches. Rich forests reappeared again and again after being wiped from the face of the planet. Nature did indeed abhor a vacuum, but repopulated the ecology with a hundred thousand species each pursuing its own ends, jostling, collaborating or competing within a multifarious biosphere.

The language of nature is written in an abundance of species. When human beings exterminate many of these species they are diminishing that language. To fail to recognise species is like being unaware of words that are essential to cogent speech. The extinctions of recent decades mean that strands in the web of communication are being slashed, leaving an impoverished language. I don't know how all those plants and animals on Ham Hill talked to one another but I could recognise that there were subtle interconnections. There is a feeling of rightness when an ecosystem is functioning as it should; the organisms are comfortable with themselves, a kind of contentment. It does not matter that competition is ruling what is happening between individuals, what matters is the sound of the language. But the names/species are the words, and without them you cannot understand the narrative. Those who don't appreciate this are effectively deaf.

*   *   *

Forge Cottage had a small greenhouse, and this became the focus for a short-lived passion for cacti. My interest was sparked by moving a miserable-looking, spiky cactus that had been refusing to die for many years into the greenhouse to see if it would cheer up. It rewarded my attention by producing a huge white flower, longer than the plant itself. The trumpet of slender, pointed petals curved outwards l ike some sort of exploding firework to surround numerous yellow stamens: it was spectacular. I started to collect as many species of succulent as I could and learned a fundamental principle of evolution. True cacti were just one family of plants that had learned to cope with arid conditions by storing water in their bloated stems, the surfaces of which became the site of photosynthesis. Other families had learned a similar trick, and several had become such close mimics of true cacti that you had to look closely to spot the differences. I would later come across many of these interesting plants in the wild when my fieldwork took me to deserts around the world, but for the moment the greenhouse was my passport to New Mexico or Namibia. Real cacti are almost confined (as natives) to the Americas. In Africa, *Euphorbia* evolved into numerous species that look very similar to cacti, but their unspectacular tiny flowers betray their true identity; no exuberant blossoms for these succulents. They are related to troublesome weeds that flourish in my garden. *Euphorbia* is one of the most widely distributed plant genera, and species can be anything from a leafy herb to a tree. All of the 2,000 species have a white juice and all the books warn that it is noxious. I can confirm that judgement: I once accidentally rubbed some of the juice into my eye and was dancing up and down in agony for hours. The ones that resemble cacti have spines, but they usually originate as outgrowths coming directly from the ribs; on true cacti spines emerge from little woolly cushions, so the two kinds of succulents can be told apart even when not in flower. There is no better example of convergent evolution than cacti and *Euphorbia*, a demonstration that similar habits and habitats generate close resemblance derived from separate ancestors, even down to details like spines. I soon discovered that much of the charm of cacti lay in the symmetry and arrangement of their spiny covering. The spines could be viciously hooked or delicately splayed, or cover the whole of the plants like a strange cobweb. I grew large prickly pears and mounds of mammillarias and ferociously armed barrel cacti (*Echinocactus*). The real triumph was persuading them to flower in a garden in a corner of the English countryside. True cactus flowers often have an odd, shiny brilliance that can almost hurt the eyes, and I had red, pink, yellow and white flowers to prove it. No cacti have blue flowers. I grew succulents that stored water in plump leaves rather than in their stems, culminating in the 'stone plants' which were reduced to a single pair of fat leaves, and really did look extraordinarily like rounded pebbles – until the surprising yellow, daisy-like flower emerged from between the leaf pair.

During my period of obsession with succulents I went back to Kew Gardens, where I had first become aware of the riches of the plant kingdom on family visits from Ealing by bus. In those early days the visit to the tropical jungle in the huge humid hothouse was the highlight of the trip, a challenge to see how long we could stay in Amazonia with its sweltering atmosphere, giant leaves, and dripping glass panes before pleading to be let out into the gardens to cool down. Then an ice cream was the highlight. But now it was straight to the succulent house, and there were all the cacti I had in small pots in my greenhouse grown to splendid maturity. Some species were true trees, with leafless branches reaching for the roof, rivals to the mighty saguaro (*Carnegia gigantea*), the giant candelabras familiar from a hundred westerns. Barrel cacti nearly as tall as I was displayed their sheaves of blades challenging all comers to breach their defences. A sign saying 'Do not touch the plants' seemed not a little redundant. Shelves were packed with mammillarias that resembled balls of wool piled on one another except that circlets of small scarlet flowers emerged from the top. The 'stone plants' had a special display at eye level, with the various *Lithops* species set among similarly coloured pebbles so that the visitor could appreciate the perfection of their disguise. I peered closely at all the labels much as I had in the Geological Museum where the fossils were on parade. I stood on the base of the railing and craned my neck to try to see some species at the back of the higher shelves. My hungry inspection was interrupted by a querulous voice: 'And what do you think you are doing, young man?' A senior gentleman with a white mane of hair and a grubby coat was eyeing me with suspicion. He rather resembled William Hartnell, the first Doctor Who. I told him that I was trying to see the names of the species at the back, and he looked very cross. Then I noticed he had a staff badge pinned to his coat that read 'Maxton'. I said with genuine awe: 'Can you be the Maxton of *Crassula maxtoni*?' The transformation was immediate. He almost blushed, impressed that I had heard of this species named after him . . . honoured of course . . . lifetime in the greenhouse . . . unsung contribution . . . He became more Dr Dolittle than Doctor Who. We talked about succulents, and I dropped what little knowledge I had into the conversation, and I had arrived somewhere I wanted to be. This was my first contact with the world 'behind the scenes', where collections are stored for scientific reference. I told Mr Maxton that this might be my ambition, and his face clouded. 'Don't do it . . . unless you want to finish up a terrible curmudgeon like me.' I had been warned.

*   *   *

The frost was kept at bay from my succulent collection in the greenhouse with the help of two old-fashioned paraffin lamps. One January night there was an exceptionally severe freeze. Worse still, one of the two lamps went out. When I realised what had happened it was too late. Many of the plants had gone all watery by the time the sun shone on the greenhouse. The poor sad things turned black and died. Only a few inhabitants of cold deserts survived; most succulents never encounter a frost in nature. I carried those few surviving plants with me for years, but I never had the heart to try to build up the collection again. Many years later in the Mojave Desert I saw barrel cacti and *Echinocereus* with its brilliant red flowers and it was like meeting up with old friends.

I passed the Cambridge University scholarship examinations. I had applied to King's College for no better reason than its famous choir and because my predecessor as head boy of Ealing Grammar School, Peter Sheldrake, had secured a place there. I found the examinations much more testing than A level, although most of them have subsequently been filed under 'best forgotten'. The physics paper had an obligatory essay question: '"What goes up must come down." Discuss.' I really had no idea how to tackle such a deceptively simple statement, and must have wittered on a bit about gravity and temperature, but I could not possibly have passed. However, the telegram that arrived telling me of my success was real enough. I had almost a year before I needed to go 'up' to university. Time to grow up.

I moved permanently to Forge Cottage, Ham, and got a job. I had previously worked briefly for Mr Lansley, the village shopkeeper and baker, delivering loaves up rutted tracks leading to remote farmhouses, the chassis of the baker's van scraping against flints and sticks to make it through. Mr Lansley's loaves were shaped rather like large bricks, and were delicious on the day they were baked, but on the following day hardened up so dramatically they could have been used to build a real wall. 'Are you the new baker's boy?' a sweet-faced farmer's wife enquired, but I was only in the job while the usual fellow was ailing. My job at Carter's was more serious. I was a builder's labourer, and the lowest of the low. Carter's was a building firm based at Inkpen, up the road, and maybe Mr Carter took me on out of curiosity to see how long I would last. The first day I had to shovel builder's waste all day, and at the end of it my hands were raw and bleeding. Austin, a kindly carpenter, took pity on me and dabbed gentian violet solution on to my wounds, which stung abominably. 'Green hands,' said Austin. Calluses formed and my hands became less green. I was paying the price for the years of evading football and cricket in the Art Room. Swinging a pickaxe all day made every muscle ache in my enfeebled intellectual's body. This was genuinely testing my theoretical socialist dedication to understanding the working class by joining it. I slowly became fitter, and a more useful labourer's mate, and less theoretical. I put on a stone's worth of pure muscle. Some genteel ladies from the village called in discreetly to tell my mother that they were concerned that her son was working with 'those rough Carter's boys'.

As we moved from job to job, the Carter's boys educated me in the slang and genetics of the network of villages around Hungerford. The local word for sexual intercourse was 'treading'. It took me a while to link this with the indecorous way drakes mounted ducks on the Kennet and Avon Canal. There was much talk about who was treading whom as we munched our lunchtime sandwiches, until the grizzled old foreman cleared his throat and muttered: 'Well, we're not 'ere to spit and cough . . .' and got us back to labouring. The mightiest Carter employee was Jim Bowley. While I staggered beneath the weight of a single sack of cement, Jim could carry one under each arm. He resembled Desperate Dan, the brawny character in the *Dandy* comic, who had a jutting, bristly chin and could lift anything. Dan was nourished on cow pie, with a pair of horns projecting from the pie crust. Jim Bowley was, they told me, a most prodigious treader. He would tread anything going. If he slipped off for half an hour when we went to Buttermere to fix a roof, not much was said, but a few significant glances indicated that some treading was going on somewhere round the back. In the surrounding villages I began to notice quite a few burly young children with jutting chins and five o'clock shadows (the latter mostly boys), and I started to appreciate that natural selection was not confined to the pages of a textbook, or to rutting stags. The Bowley genes were a-spreadin', even without the help of cow pie.

The communist in the adjacent village of Shalbourne bought me a beer in the village pub. Mr Wiggins the pub landlord made it obvious that he would have preferred me to consume it in the saloon bar along with the other people with middle-class accents. The communist thought I was a class warrior. If Miss Simkins, the village gossip, had seen me she would have said it was bad enough that this young man was consorting with those rough Carter boys, but now he was conspiring with a known Bolshevik. She would probably have taken a more benign view of my other efforts to integrate with local life. I joined a drama group. The problem was that I was just about the only male to volunteer, and certainly the only young one. The director had to find a play with many mature female roles and very few male ones, and especially only one romantic lead. Such a play did exist: *Great Day* by Lesley Storm (1943). I was the romantic lead. It concerns the misadventures that accompanied a visit by Eleanor Roosevelt to the Women's Institute in a small village during the war.*The director said it was a 'period piece'. Numerous women of a certain age got into comical scrapes; stage effects were enhanced by the inclusion of a few farmyard

---

* It was based on a true event, the visit of the president's wife to Barham Women's Institute in 1942.

animals like ducks, rabbits and hens in cages. The only other male was the 'caretaker' who had to do the business with a broom from time to time, while uttering comical asides. The romantic lead was on leave from the front, so I was dressed as a soldier, and shared some slightly excruciating scenes with Daphne, my beloved, somewhat along the lines of *Brief Encounter*. Daphne was a young girl from Inkpen, and when I uttered lines like 'Darling, I can't live without you,' she really thought I meant it. Two performances were given: the first in Inkpen Village Hall, followed by its equivalent in Ham. During the second performance I was just trying to inject the right dose of sincerity into the line 'Never mind, darling, this too, too beastly war will soon be over' when one of the chickens laid a particularly large egg, and celebrated the event with a bout of enthusiastic clucking. It brought the house down, and my theatrical career to an end.

While I worked at Carter's I met my first proper girlfriend, the daughter of the local vet in Hungerford. My previous amours had all been crippled by my diffidence. One of my sister's friends, a local farmer's daughter, had captured my heart completely, but all I did about it was to initiate interminable telephone conversations, and somehow never manage to ask her out. Worse, I fell in love with the girlfriend of one of my closest school friends, the brilliant Krzystof Jastrzembski. His father had fled Poland during the uprising against the communist regime in 1956, when many Poles settled in Ealing. They were welcomed: the bravery of Polish airmen during the war was a recent memory. Krzystof's father had been something in the government, and could not remain in his homeland. Kris (as he became) was clever, tall, dark and very handsome, with curled locks like those of the Discobolus of Myron. He resembled the actor Zbigniew Cybulski, who starred in the wonderful films of Andrzej Wajda (*Ashes and Diamonds*) that were emblematic of Poland's cultural awakening at the time. Martine was blonde and French and to me impossibly beautiful, and for that reason obviously belonged with the comely Pole. A spindly, spotty, cerebral person could aspire to the position of best friend and confidant, and perhaps even secretly revel in the suffering of an unrequited admirer. 'Alone, and palely loitering' seemed to be the part for which I was best suited. Had I confessed to my mother (heaven forfend!) she would have countered with a brusque 'faint heart never won fair lady' and my heart was, truly, not a little faint. It was odd how a spell as a builder's labourer beefed up the heart as well as the muscles.

It would be disingenuous to pretend that my time at Carter's was a political gesture, in spite of my being flattered by the village communist. I was saving my modest wages and boarding at home cost-free; I was not living on the money. However, I did now better understand the implications of being a member of Marx's proletariat. One morning I arrived five minutes late at Carter's yard after sustaining some damage to my bike – no excuses, I was sent home by Mr Carter himself and docked that day's wages. If the Carter crew were 'rough boys' it was because their lives were tough. The one advantage was that when the day's work was over both body and mind were free:

there was no room for brooding. I was fully relaxed for the first time: A. Sainsbury-Hicks was history. The intense intellectual of my teenage years was fading, and only now can I mourn his passing; that intensity of focus, that breadth of ambition. Enthusiasm for natural history was not starved: I continued to add to my collection of fossils at the weekends. After I learned to drive I used an ancient Austin 10 (registration number OW 6686) to reach chalk quarries at Ogbourne St George, near Marlborough, where rare examples of extinct snails, and even a coral or two could be collected. On Saturdays, a pub crawl with the local river-keeper's son was a fixture, and I should never have driven the Austin back from Marlborough along such tiny country lanes, bouncing occasionally off the high banks. Bucolic regularly, alcoholic occasionally, melancholic rarely: it was a restorative interlude.

The money I earned at Carter's was spent on an adventure travelling to Morocco with Kris Jastrzembski. At that time hitch-hiking was customary, supplemented by bus journeys where necessary, and we spent a long time travelling southwards across France and along the southern coast of Spain. We stopped for a break near Almeria, at the edge of the desert in the Cabo del Gata, the hottest and driest part of Spain. A seaside village called Carboneras was then almost as it had been for centuries: low, square houses with thick walls and small windows and everything whitewashed. Old widows sat in doorways, dressed in black from head to foot; they may have lost their husbands at sea long ago. It was very quiet. A few pesetas went a long way then, and we were able to stay in a clean room in a small *pensión* for the cost of a cheap supper back in England. Generalissimo Franco was still in charge, and the old Falangist emblem of yoke and arrows, somewhat battered now, greeted the traveller as the bus entered the village. A member of the Guardia Civil wearing a curiously flattened, shiny black hat was there to keep order, but he didn't have a lot to do. Part of the movie *Lawrence of Arabia* had been filmed there, and a dry riverbed nearby still held faded remains of cardboard palm trees that had once sheltered David Lean from the relentless glare.

After the trance of sweltering Spain, crossing from Algeciras by the Straits of Gibraltar to Tangiers was a profound shock. My first baptism into another culture was more than a splash on the forehead – it was total immersion. The music was different, the smells challenged identification, and the crowds of men in their djellabas seemed frantically bent on mysterious but urgent business. The narrow streets in the medina were piled with colour: swatches of garish fabrics vied with sacks of deep red paprika, yellow turmeric, brown cumin, glistening dates. Artisans showed their skills to passers-by, beating copper into shape, working leather, or boiling sugary snacks. It was like travelling into the bustling maze of a medieval city, where all the trades were still on open display. Occasionally, a highly decorated door indicated the entrance to some old and grand private courtyard. We found the Café of the Dancing Boys where it was clear that this urban society acknowledged homosexuality without public censure. There was an ill-lit room at the back where *kif* – cannabis resin – was smoked in tiny clay pipes by middle-aged hashish-lovers. We became accustomed to a characteristic sharp tang in the atmosphere as we took off southwards.

Hitch-hiking proved riskier in Morocco than in Europe. No doubt Kris's handsome features attracted attention. One last lift in an opulent Mercedes took us along a very dusty track. The driver was dressed in a fine, silky, pale blue djellaba and a selection of rings sparkled on his fingers. He was a sheik and would take us to his palace in the desert – no, he could not possibly accept our excuses. He smiled broadly, revealing gold. We swept into the courtyard of an extensive single-storey building, and were taken to a spacious reception room surrounded by a ledge bearing many plush cushions. Intricate tiles all interlocking curlicues and arabesques made up the floor. The obligatory mint tea was proffered – fresh mint leaves, much sugar from a solid block, steeped in boiling water, and served in a glass (how tired we became of the quip 'berber whisky'). Mr Marney's French was very useful. Since both Moroccan host and his English guests spoke inadequate French we conversed quite fluently. Couscous was served. We remembered to rinse our hands in the rose water placed in a bowl for the purpose, and to use our right hand for eating. Gold smiles flashed, particularly in the direction of Kris. When we said we really must be on our way we were told that it was quite out of the question, and that we must stay as guests. We were shown to a simple room with comfortable-looking beds and a jug and basin in the corner. After waiting for about half an hour we tried the door. It was locked from the outside. We did not have to engage in much discussion before opening the small window, checking that our packs would fit through it followed by our lissome bodies falling uncomfortably to the ground; we headed off in the dark to the track that had brought us to the sheik's redoubt. We were not pursued. Once back on the main drag we picked up an honest lift as far as Marrakech, where we met up with a Norwegian boy with considerably more money than we had, and together we bought an old Citroën 2CV. This flimsy contraption – as much tin can as automobile – took us over the Atlas Mountains by way of the pass of Tizi-n-Test. I shall never forget the extraordinary views south to the fringes of the Sahara Desert. Not far away, there were groves of ancient cedars on the high slopes, whispering gently in the wind.

The Anti-Atlas to the south was more like a series of hills in the desert. Everywhere was so dry; parched plains between the hills relieved by a few thorny bushes eking out a meagre existence waiting for the first drops of rain. Strata were beautifully displayed on the hillsides, the geology written on the ground in differently coloured layers as if designed to share the narrative of the rocks with every passing traveller. I was not able to persuade my companions that a day or two walking up a wadi would be time well spent. The 2CV choked on the sand, until we kept her going by attaching strings to the toggle controlling the petrol supply.

At that time there were no small hawkers trying to sell fossils. The locals did not yet realise they were sitting on a palaeontological gold mine. When I returned several decades later the hills around Erfoud and Alnif had become the source of trilobites for hundreds of rock shops around the world. Every small bush hid a smaller boy. No matter how far off the main road you were, as you approached in your vehicle the boy would emerge from cover with outstretched hand: 'Trilobite mister? Good price specially for you.' I traversed the Anti-Atlas in search of rare species, but by now the fossils had acquired a commercial value, and the desert was pockmarked with excavations made by the native Berber folk in pursuit of valuable rarities. Local entrepreneurs had become expert in extracting beautiful (or even rather bizarre) trilobite species from their rocky redoubts, and some of these middlemen had made a lot of money. They travelled to 'fossil fairs' around the world to sell top rarities to serious collectors with deep pockets. It was a wonder those Berber workmen could labour under the relentless sun, smashing rocks, but by the time of my last visit with Sir David Attenborough in 2010 the television people were able to summon dozens of extras to belabour the hillsides with pickaxes and hammers to make a perfect background for a piece to camera. I was well known for my trilobite research by then, and when we visited Alnif my arrival caused some excitement. Fossil merchants wanted to be photographed with me alongside: this was the only place on earth where I was better known than David Attenborough. By then, many of the local rock shops also sold fake trilobites manufactured in their dozens from moulds vaguely based on real specimens; a dollar or two for a piece of bogus ancient history.

My first trip to the land of the trilobites did not end

well. I picked up dysentery, not the kind that is over in a day, but some other persistent microbe that made my life progressively miserable. It was a challenge to ingest enough fluids to stop dehydration. With my growing beard and dressed in my own brown djellaba, and sandals manufactured from old car tyres, I looked like an ailing Moroccan whose days were numbered. Kris and the Norwegian managed to find an expatriate German (implausibly known as Desert Jim) in Marrakech to take the 2CV off our hands for cash. We may not have owned the car legitimately, or so Desert Jim said, so it was whisked off immediately and broken up into parts for reuse. There was just enough money to fly home from Tangiers. This was my first time in the air, and it should have been thrilling, but all I remember was astonishment at being able to look down upon clouds. When I arrived back in England my mother failed to recognise this cadaverous figure with staring eyes. Even in my diseased condition I enjoyed making such a dramatic entrance. An emergency trip to the Radcliffe Infirmary in Oxford got me the sulphonamide drugs I needed. The sad thing was that all the muscles I had built up during my time at Carter's had just wasted away. The best bits of my body had vanished to feed an unpleasant germ. I was gangly again. It was the price paid for a measure of growing up.

*   *   *

Now, I was bound for Cambridge University to read natural sciences, so that should be an end to the intellectual blunderbuss. Time to buckle down and find a single target. In October 1965 my belongings were packed into a locked trunk, and placed on the 'Varsity' railway line that then connected Oxford with Cambridge (it is being reopened in the twenty-first century). I arrived with the trunk outside King's College at the centre of the ancient city. Visitors had to go through the gatehouse to gain access to the inner courts. The college porter sitting grandly in his office looked down a list, nodded, and then addressed me as 'sir', something that had never happened to me before. My trunk and I were directed to Market Hostel, where I spent my first year. Across the road from the college on the other side of King's Parade, my small, modernised room commanded a view of the old market square, set out with stalls selling fruit and cheese, with a selection of less conventional hawkers shouting from booths. Great St Mary's Church was around the corner, where the vicar of Cambridge, Canon Sebag Montefiore, delivered his famous sermons (I never heard them). Nothing can prepare a young man from a London grammar school for the beauty of King's College, and while you are a resident member of the college it is, however briefly, yours. Life in the college becomes as important as life in the specialist faculty of economics, science, languages, or whatever. Unlike many colleges, with their tight cloisters and secret passages, King's is all air and space. The chapel is a supreme architectural achievement from the end of the medieval period, despite one of the dons describing it as a sow on its back. I was at the college long before women were admitted, and there was even a midnight curfew when the grand front entrance was locked. During our tour around the college we were shown (discreetly) how stranded undergraduates could climb in around the iron back gate, by shuffling above the muddy ditch that bounded the grounds. An intoxicated tumble into the duckweed was one of the necessary initiations.

Until I arrived in 1965, I had no idea that King's College was the traditional haven for homosexuals, a place where gay men bound for Cambridge from their public schools could be comfortable in the company of their friends, without attracting the attention of the law.* Several of the senior dons dated from that time. I briefly met E. M. Forster, by now a kindly old gentleman who spent his last days in the college as an honoured resident. The ex-provost Sir John Sheppard was nearly as old, suffered from dementia, and was prone to wandering around the quadrangle looking vaguely for whichever Greek boy had taken his fancy. The senior tutor, the Milton scholar John Broadbent, set about admitting more state-school students, and I doubt that the college had ever seen a pair of blue jeans before his appointment. Younger dons tended to be radical, and heterosexual, like most of the new generation of undergraduates. It was an interesting period of transition in the life of the college. For a former employee of Carter's the builders it was extraordinary to hear the languid drawl of Etonians for the first time, emanating from one of the younger Mosleys or his friend Nigel Honorius Sitwell.

---

* Homosexual acts between consenting persons were not decriminalised until 1967, two years after I went up to King's.

Rugbeians, Salopians and Wykehamists were hardly less exotic, as expensive private schools sent their brightest onwards to King's. It was almost as alien to me as my first immersion in a Moroccan souk. I realised that for these self-confident undergraduates a place in the splendid old college was taken for granted, just the next step in a privileged life. They probably felt completely at home with its haughty grandeur. They had been accustomed to portraits of distinguished predecessors in gowns and periwigs looking down from gilt frames. The perpetuation of the class system became clear to me. I now knew who would get the best-paid jobs, so that they could afford the education to get their children the best-paid jobs. I became what my mother would have called 'chippy'. It took me some years before I could see past the privilege to the person.

By contrast, one of the great virtues of science is that it is truly egalitarian. Whether you talk with an Estuary twang, flat Midlands vowels, or were brought up to say that you lived in a big hice with extensive grinds, it simply does not matter in the laboratory. It was not always so. Many of the famous scientists of the eighteenth and nineteenth centuries came from aristocratic or wealthy backgrounds – the Darwins or the Cavendishes come to mind. Educated clergyman/scientists were equally not 'trade', let alone of the labouring class. In the mid twentieth century merit alone finally became sufficient for employment, and women earned proper recognition at last. Some of the most brilliant scientists I know have had a much harder route to science than mine, do not talk 'posh', and rightly don't give a damn about the niceties of cutlery usage at official functions. It is surely an advance in culture when the question 'where did you go to school?' can be replaced by 'what have you discovered?'

A particular virtue of the Cambridge system was its flexibility, which suited a butterfly like me. Within the broad natural sciences tripos students could change their principal subject as they better identified what they wanted to become. Three subjects were obligatory from the first, Michaelmas (autumn) term. The snag was that there were examinations at the end of each year, so there was no fooling about for three years until all-important finals loomed, as happened in Oxford. This system may have engendered Cambridge's reputation for seriousness; equally, it might explain why Oxford produces so many politicians. I was able to study biology formally for the first time, even though I had been a biologist manqué since I could remember. I said goodbye to chemistry, but my training in the properties of the elements did not go to waste. Geological science was then rather absurdly split into two departments: geology itself and mineralogy and petrology, known to the faculty as 'min and pet' like the names of a pair of cockatoos. The departments were housed in slightly forbidding buildings, including the Victorian Sedgwick Museum of Geology on Downing Street; each department had its own professor. This is where I would spend much of the next five years.

College life dominated my early days in Cambridge. I made friends very quickly, and one or two of my fellow undergraduates have remained friends for life. Nearly all my new intimates had come to Cambridge by a route similar to my own. As my mother was a widow with a low income I was supported by a full government grant. I felt affluent enough to smoke too many of the smallest cigarette known to man, Player's No. 6, which a friend from Nottingham (where they were manufactured) said were made from the sweepings off the factory floor. It suited me that colleges had nothing to do with the university departments or faculties, so I was surrounded by students reading English, economics or medicine. There was no other King's geologist in my year of matriculation. Although my polymathic ambitions had scaled down, I was still writing seriously all the time, and sacred music in the chapel introduced me to the glories of polyphony. I slipped into the back of the room to hear lectures by Raymond Williams on Marx and literature. F. R. Leavis was still in charge of his many acolytes. Some part of me still wanted to be living proof that the 'two cultures' – arts and science – which C. P. Snow had promulgated in 1959, could be embodied within a single cultural life. I was loosening up. My friend Victor Gray introduced me to jazz, particularly Miles Davis and *Kind of Blue*. I heard Thelonious Monk at the Students' Union. I followed the Beatles like everyone else. I talked too much and drank too much. I do not think the fifteen-year-old avant-gardiste would have approved the new model, but this one was probably better company.

8

Three Lobes

The trilobite on my desk came from Morocco. When I was travelling with Kris Jastrzembski I must have passed close to its original home in the Anti-Atlas, where the hills are made of limestone rock beds almost 400 million years old – but I did not know that then. This trilobite is my own messenger from deep time: a Devonian Hermes. As I write, the trilobite gazes at me through a pair of prominent eyes, a stony stare if ever there was one. The eyes are so well preserved that I can make out files of small, convex lenses lined up along the surface of each one. If I stare at the trilobite, he stares back. The compound, faceted structure of the eyes quickly proves that the fossil must belong within the arthropods, the most diverse group of all the animals in the biosphere; countless arthropod species have crawled on their jointed legs over every metre of the planet, from high mountains to the abyssal depths of the sea. Insects, crustaceans and spiders are all arthropods – and so are (I should say were) trilobites. Eyes are invariably positioned upon the heads of arthropods, so the front part of my trilobite, much of it taken up by an inflated lobe between the eyes, must also be where its brain was situated. If it were alive, the trilobite would crawl head first across my desk on the many legs concealed beneath its carapace and land in my lap, scrabbling. It would make quite an impact, too, for the specimen is at least six inches long. It was preserved as a fossil because its upper surface was protected by a thick 'shell' of calcite, something like that of a lobster, which is hard enough to fossilise very well. Now, that carapace is black, like the limestone that encloses it – but who knows what colours the trilobite

The large Devonian trilobite *Drotops* from Morocco.

may have had when alive upon the Devonian sea floor? It might have been as gay as a clownfish: sadly, pigments very rarely fossilise. Behind the head, an originally flexible thorax comprising eleven, similar-looking segments separates the front shield of the animal from a tail (pygidium) at the back composed of another nine or so segments all fused together. The 'three lobes' that give the trilobite its name run from front to back, as the middle lobe of the animal is elevated above those that flank it on either side. The whole upper surface feels rough to the touch, because it is covered with large tubercles. This was a businesslike arthropod, a well-protected army tank of an animal that had to be taken seriously in the Devonian ocean. It was no primitive dullard but a creature of its times, adapted to its own world, as closely as a skateboarder to the streets of New York. It has a scientific name, *Drotops megalomanicus*, that was given to it as recently as 1990 by the German palaeontologist, Wolfgang Struve. He clearly wanted its name to acknowledge this species' unusual size among its contemporaries. When I passed through the hills of the Anti-Atlas all those years ago my trilobite must have been lying concealed within the strata, just waiting to be discovered.

This megalomaniacal *Drotops* is a special object that was acquired more recently than the trout that began this book, or the Jurassic ammonite, my very first fossil. I was not permitted to keep a personal trilobite collection during the many years I worked at the Natural History Museum in London; the director considered that there might be issues around conflict of interest, although my own acquisitive instincts were fully satisfied by working with the national collections on a daily basis. *Drotops* was a gift from my publisher when my big book about evolutionary history (*Life: An Unauthorised Biography*) was published in 1997. The fossil's robustness makes it an ideal specimen to take as a 'show and tell' when I talk to schoolchildren about life in a great museum (I sometimes have to ask for it back). I received another trilobite gift that could have been equally exemplary. This one is much smaller – the size of a large marble – and it is tightly rolled up into a ball, a feat accomplished by many trilobites thanks to the perfect articulation of their thoracic segments, which can glide past one another like the plates on the joints of a medieval suit of armour. This specimen is older than *Drotops* (by about 30 million years), and is a beautiful example of the 'Dudley bug' *Calymene* from the Midlands of England; it was left to me in 1999 by Sir James Stubblefield – the grand old man of trilobites – when he died aged ninety-eight. It came with a fragile letter, revealing that the *Calymene* had originally belonged to Professor Hawkins of Reading University, who willed it to Sir James. And Professor Hawkins had himself received it as a talisman to protect him in the trenches of the Great War. The letter from the donor to Professor Hawkins expressed the hope that the armoured 'bug' would offer an appropriate kind of luck to help its new owner survive the ordeal. Herbert Leader Hawkins did indeed survive and prosper, to pass the trilobite down the chain that led eventually to my mantelpiece.

Another chain of circumstances tied my own future to trilobites in ways I could never have imagined when I took up my place in King's College. I did know quite a lot about the extinct arthropods, ever since finding my first examples on the cliffs near St David's. They soon became the favourites in my collection. The idea that they could ever become attached to a pay cheque would have seemed inconceivable. The events that eventually led to such an outcome illustrate that my good fortune was as considerable as that needed to survive the Battle of the Somme. What my mother always called 'the professions' had defined, clear routes to an income. Medical doctors know more or less how they will emerge at the end of their training. My own trajectory was unique. I wish I could claim that my route to science was the consequence of a particular set of virtues, but I would rather believe in a supernatural Fate with a quirky sense of humour and a penchant for rolling dice to see what happens next.

There were false starts and red herrings. The idea of at last reading biology was enticing, and began well. My first college tutor Dr George Salt was everything an old-fashioned don should be – courteous and learned, and without a hint of patronisation. He invited me to tea at his home, a dark Victorian villa that reminded me of my visit to Mr Morley-Jones in his study. A grandfather clock ticked in the background as the tea and cake was brought out on a proper service. I found myself sitting rather too upright on a slightly uncomfortable chair. My host's wife called her husband 'Salty'. Conversation was a little stilted, but not unfriendly. Later, I spent a long time writing my first essay on the geological history of insects, and if it was just a precis of the reading I had done with nothing original about it Dr Salt was too kind to say so. I did not do quite so well with the college botanist, who I was hoping would be impressed by my work on the chalk flora of the Berkshire Downs; it was disappointing when he failed to register much interest. I wondered why: perhaps too much naïve enthusiasm was not quite the thing; I had to learn to be cool. I did not fare better on a fungus foray led by the famous mycologist Professor Corner. When I emerged from the bushes with my basket and pronounced I had found *Clitocybe nebularis* (which I had) or *Coprinus comatus* (ditto) he looked at me with a measure of distaste. I think it may have been *his* job to pronounce. As the day went on, I observed that his expression became like the grimace walkers direct at enthusiastic Labradors that insist on repeatedly returning with sticks, wagging their tails. They somehow exceed the bounds of propriety. I was disappointed, mostly with myself, and the rules I had failed to appreciate, but also with the don who might have recognised a devotion to natural history that at least deserved a pat on the head, which is all that a Labrador needs to carry on being irritating.

Genetics was new ground for me. Nowadays, I assume university teaching begins with DNA and gene sequencing and goes on from there, but when I started out the bread and butter of the course was classical Mendelian inheritance, understanding the expression of dominant and recessive genes, and the organisation, structure and function of chromosomes. This was way off my countryside interests, but good stuff to know. Whereas Gregor Mendel originally used plants to understand the principles of inheritance, by the middle of the last century the small 'fruit fly' *Drosophila melanogaster* had become the experimental workhorse for genetics. *Drosophila* was particularly prone to mutations of features like eye colour or wing development that could easily be observed under a binocular microscope. In the wild, most of these changes are rapidly eliminated as they reduce the fitness of the flies, but in the laboratory they are a boon to understanding how, and in what proportions, genetic mutations are handed on from one generation to the next – and it takes only a couple of weeks to breed up the next experiment. In the examination at the end of my first year we were given phials containing living flies that we had to kill with some liquid chemical so that we could examine the population and determine the numbers and ratios of mutants in the samples. I added whatever-it-was and the flies all dropped to the floor of the tube where they lay, immobile. I tipped them out on to a Petri dish and started counting. I was horrified to see that after a minute some of the flies started twitching. They were waking up. A number of them flew away – never to be seen again. I must have failed to administer enough dope. I am ashamed to admit that I made up the results for the examination paper – I guessed at what they wanted us to prove. That was the moment I realised the advantage of working on animals that had been dead for a very long time and could not fly away. Palaeontology was the intersection between geology and biology, the best of both worlds after all. If I had briefly considered being a geneticist, that fantasy flew away with the flies.

The Cambridge terms lasted for only eight weeks, but were extraordinarily intense. Would-be scientists had to work more hours than students of the humanities. Not only were lectures obligatory in the mornings, but laboratory work also took up the afternoons, and essays had to be written in the evenings. Only the medics worked harder, learning in their first year the endless roll call of our bones and organs. I could hear them in adjacent rooms intoning aloud the details of human anatomy to drum it home – 'organ recitals' I called them. There was something annoying about watching my friends who read English, still in their pyjamas at eleven o'clock in the morning nursing their hangovers when I had already been at work for a couple of hours. I invented a kind of automatic writing so that I could take notes during a nine o'clock lecture regardless of my condition. It would serve me well in committee meetings in the years ahead. Generally, I was not happy during what Americans call the freshman year. I experienced delayed grief for my father; I was still confused about my scientific destiny, and my attempts at being a Renaissance man were faltering. At Ealing Grammar School for Boys I had been top dog, and here I was a nobody. Even a mycological professor found me tedious. I had taken to reading about Eastern philosophy. Alan W. Watts's books explaining Zen Buddhism helped, or maybe confused my attempts to discover whether there were 'truths too deep for physicists' as my poem had it. Depression descended upon me like an incapacitating blanket. For a short while I was under medical supervision, and took pills that numbed me as much as helped me. I was comparatively lucky: one of my fellow students never emerged from a cloud of confusion, and had to leave. It seemed to me that the Old Etonians and Wykehamists in King's were immune to such angst: they had had an expensive inoculation against it. Somehow, I continued to churn out the necessary essays on my Brother Deluxe typewriter. *J'ai survécu le tremblement de terre*, as my old French textbook would have said: I have survived the earthquake.

I also discovered debt. King's College had a buttery that was open in the evenings, where the 'gentlemen' could furnish themselves with their perquisites. It was a secret storeroom of delicious luxuries, staffed by the college butler in evening dress. The catch for a simpleton like me was that everything went on a tab to be settled at the end of term. Those Turkish cigarettes were a magnificent improvement upon Player's No. 6 – why, they even had an elliptical cross section! 'Several young gentlemen have remarked that the Chambertin '59 is very drinkable at the moment, sir,' the butler said, encouragingly. 'In that case I had better take a couple of bottles,' was my response, while the butler made notes in a kind of ledger. So it continued for a whole term. I must have learned a great deal about good wine. When the buttery bill arrived at the end of Michaelmas term I was aghast. Could I have really spent that much? My embarrassment was exacerbated by the knowledge that all my university fees were covered by a full government grant, as my mother had little income as a widow. Here was this freeloader, guzzling Chambertin; I deserved to be branded as a debtor. Then I remembered that I had tucked away a small legacy that came directly from Granny Fortey. It very nearly covered the whole of the buttery bill – my entire inheritance gone in a trice. I steered clear of temptation for the rest of my time at King's College. I became frugal. From then on it was back to the cheap cigarettes and Blue Nun Liebfraumilch.

During the first Easter break the geology course took the undergraduates to the Isle of Arran, lying a short distance off the west coast of Scotland; Arran was the traditional nursery for the science. To save money I hitch-hiked northwards up the Great North Road (A1), attempting conversation with a Glaswegian trucker with an accent more impenetrable than ancient Icelandic. I arrived at the ferry just in time. Arran is a geological much-in-little, with the main classes of ancient rocks – igneous, sedimentary, metamorphic – all displayed in outcrop, and the whole island decorated with relics of the last great Ice Age. The considerable granite mass of Goat Fell dominates the topography. Traditionally geological hotels welcomed students to Brodick, the only town of any size, and they were glad of the out-of-season trade. For a week a bunch of us followed W. B. Harland, our lecturer, as he demonstrated contacts between one rock formation and another, or showed how the hot granite arising from deep in the earth had baked and altered the 'country rocks' that it intruded. We followed the Highland Boundary Fault through a pretty glen, and failed to find fossils in black shale exposed in a streambed there, as generations of students had failed before us. I loved it all, revelling in those steps from observation to understanding I had first enjoyed with Mr Williams. My only problem was that W. B. Harland was very

fit, and shot up hillsides like a mountain goat, so I sometimes trailed in at the end of his explanations. The final
ascent of Goat Fell was my worst performance. I had
foolishly decided to wear my Moroccan sandals, with tough
soles made from old rubber tyres, but with poor grip and
no protection from the elements. The weather turned
wintry, punctuated with snow flurries. I lagged further
behind, watching Harland's nimble, stick-like figure ever
further away, only pausing from time to time for a bit of
extravagant gesticulation. My feet froze. I fell in with the
only other fledgling geologist as ill equipped, a small
woman from Barbados who had never seen snow before.
Together, we limped back at the end of the day, not much
wiser about granite. W. B. Harland was later to be crucial
to the next spin of the dice that led me on to trilobites.

I was happier in my second year. I had begun to enjoy
geology,* and my essays began to show some independence
of thought. For years, I had been trained in the regurgita-
tion of facts in the interests of good grades, and never
really understood that learning was about ploughing a new
furrow. My verse was being published in a slim magazine
called *Pawn* (I have to be shy about saying that my writing
career was launched in Pawn). It cost one shilling and
sixpence, and there was an image of a chess piece on the
front cover. I was published alongside others who subsequently became widely known – Clive James, Dick Davis
and Clive Wilmer, as well as Robert Wells, so I must have

---

* I have mentioned my second-year field trip back to western Wales in
Chapter 4, when I discovered my talent for finding fossils.

made some artistic progress. I have kept this magazine all my life as a voucher of a life that might have been.

I made one last deviation towards a wider intellectual compass. One of the courses available as an addition to the regular tripos was in history and philosophy of science, which had a small department tucked inside the Whipple Museum – a strange and compelling collection of astrolabes and early clocks in Free School Lane. The required reading could not have been more different from the earthly realities of geology and mineralogy. The debate about what distinguished the scientific method from metaphysics was in full swing, so I read Karl Popper at an early stage, and understood that science was as much about presenting hypotheses that were capable of being disproved as it was about establishing universal truths. I began to see Charles Darwin as almost inevitable – rather than as a lone genius – through the writing of Thomas Kuhn, who demonstrated that apparently 'new' ideas had their appropriate time in history, regardless of the name that carried the banner for change. I learned about the alchemical roots of modern science. I struggled briefly with symbolic logic. This new discursive world of letters fought it out with Zen Buddhism in some subconscious boxing ring deep in my psyche. Was enlightenment compatible with the Enlightenment? I learned that the incomparable Isaac Newton seemed to be able to encompass reason and religion inside one great brain – but not happily. The history and philosophy course allowed me to explore style in my essays as much as content, although I found puns hard to resist, as in 'he always put Descartes before the horse', although the context in which this astonishing aperçu was employed has escaped me. I was having such an interesting time that I did wonder whether to become a historian or critic of science rather than a practitioner. Such a transfer would have been possible under the Cambridge system, and might have laid the ghost of A. Sainsbury-Hicks's fundamental question about 'people or things'. One session of supervision with the brilliant philosopher/historian Nicholas Jardine may have nudged me back to natural science. I had worked hard upon my essay, but still recall the young don's assessment that my work was no more than a 'well-written precis of the recommended texts' – not so much damning with faint praise, as faint praise mitigating a decisive put-down. I was to discover that such remarks were typical of academic life, and competitive Cambridge academic life in particular, but it made me cautious about entering that particular arena. If that were the weaponry I would prefer to fight elsewhere.

The tutorial system was, and still is, touted as the great advantage of the Cambridge collegiate system. The image is of a kindly don offering wisdom and constructive criticism as crumpets toast before the fire, while a few, select students sit in comfortable armchairs to receive aphorisms they will treasure for life. The nearest approach to this ideal that I experienced was just one tutorial on Charles Darwin hosted by a Jesuit priest. The philosophy tutorials – not least Dr Jardine's – did serve to keep me on my toes, but were hardly cosy. The ones in 'min and pet' were often discouraging, but I greatly enjoyed the practical side of the study of rocks (petrology). This involved peering through a microscope to examine sections through samples ground thin enough to observe the optical properties of the individual mineral grains.They came in all colours and made beautiful patterns: mosaics or swirls that changed again in polarised light.The department was a pioneer in sophisticated microscopy. I learned forensic detective work that led to the identity and chemistry of the many rock-forming minerals, and my old enthusiasm for the dance of the elements was rekindled. The lectures were comparatively dull. The 'prof' was W. A. Deer, whose great work on rock-forming minerals was co-authored by R. Howie and J. Zussman; his lectures were not much more than a trot through the book. His tutorials came alive only if his seminal work on the Skaergaard Intrusion of Greenland was mentioned.This layered igneous mass was where Deer deduced many important facts about the evolution of magma chambers, but the adventure of fieldwork in a remote part of the world lifted it from being a recitation of scientific results.Who would not want such a pioneering life? A second tutor in 'min and pet', Ian D. Muir, was an expert on basalts who seemed totally unenthusiastic about the job of guiding the young; indeed, on one occasion he nodded off in the middle of our supervision. Geology was better. Peter Friend made sedimentary rocks easy to interpret, effortlessly explained their origins, and he, too, had stories of working in Greenland. The head of department when I first arrived in Cambridge was Oliver Meredith Boone Bulman, the world authority on fossil graptolites, and he was a formal, very old-fashioned style of professor, but a great lecturer.The real treat was a seminar from the curator of the Sedgwick Museum, Bertie Brighton, who made palaeontology thrilling with his vivid explanations of how you could infer much of the living animal from the ancient remains preserved in rocks. I still believe that reanimating life that has passed from the earth is one of the most interesting things to do as a palaeontologist. Through him I learned that geological time is written in the succession of fossils – each time interval with its own diagnostic fauna (or flora). Learning the succession of British fossils was routine, even a chore, but I am grateful now that I persisted with the task. Even today as I travel through the geological landscape it is animated in my mind by tableaux of the creatures that once lived there. My knowledge is transformative: a comparatively banal geological map becomes a visualisation of the history of life.

When I was offered the chance of my own Arctic adventure I signed up immediately. For some years W. B. Harland had been running expeditions to Svalbard, a group of islands way north of the Arctic Circle. He was rapidly becoming the authority on the geology of the largest of the islands, Spitsbergen, somewhat to the annoyance of some Norwegian geologists. After all, they had first claim on the Svalbard group, which was administered by Norway's own *sysselmann* (governor) – although other countries had rights there, too, including the United Kingdom and Russia. Harland's new expedition was to take place during the long summer vacation at the end of my second year. In the previous summer of 1966 one of the boat parties returning to base had stopped off along Hinlopen Strait adjacent to the most remote northern part of the island. They wanted to collect fresh drinking water from a melt stream running across a raised beach in front of the huge Valhallfonna glacier that occupied the centre of Ny-Friesland, as this part of Spitsbergen is called. The landing party was surprised to discover fossils littering the foreshore. No fossil-bearing strata had been mapped in this area before – this was undoubtedly an exciting new discovery. A small collection was made on the spot, but there was no time for a proper investigation. Trilobites were conspicuous among the small bagful of specimens that were brought back to Cambridge.

By another curious twist, Professor Bulman's tenure as the Woodwardian Professor of Geology had come to an end during my first year, and his replacement, Harry B. Whittington, was an authority on trilobites. He had been recruited from Harvard University. It was rumoured that his rival for this famous post – the equally distinguished brachiopod researcher Alwyn Williams – had queered his pitch by actually applying for the job. The Cambridge way was to issue an invitation, not to have somebody ringing the doorbell like a tradesman. The Woodwardian Professor's office was at one end of the Sedgwick Museum, a classic nineteenth-century treasury lined with ranks of polished, glass-topped mahogany cabinets full of fossils laid out in rows. Approaching the professor's redoubt through the gallery was like drawing near to the altar after progressing along a nave enclosed by holy relics. The destination was impressive, and not a little awe-inspiring. Visitors tapped on the door inscribed in gold letters 'Woodwardian Professor' with their hearts in their mouths. However, Whittington could not have been more different from the aloof Professor Bulman. He was dark-haired, with a neat moustache, and many years spent in the USA had concealed the flattened vowels of his Birmingham accent. He came with American informality. He insisted on Christian names, which Bulman reserved for his wife and children, and that only behind closed doors. When the fossils from the chance Spitsbergen discovery were shown to 'Harry B' he pronounced quickly that they indicated mid-Ordovician – the first rocks of this age (465 million years) known from the whole Svalbard Archipelago. Some of the trilobites were familiar from species he knew well from Canada, and especially Newfoundland, where they could be collected from similar-looking dark limestone. Clearly, this part of Spitsbergen demanded a proper geological investigation. A dedicated visit to Hinlopen Strait was planned as part of the 1967 expedition. A final-year student, Geoff Vallance, would be the lead geologist, and he needed a field assistant from the second year. That would be my job.

I used this expedition to Spitsbergen to introduce the concept of geological time in a book written more than twenty years ago. This trip also played a crucial part in my journey to science. If I had chosen to do something else in the summer of 1967 my life would have been very different. I only have to hear the Beatles' anthem 'All You Need is Love' to be transported to a vast shingle beach at nearly 80 degrees north, where our tough Whymper tent hunkered down among the cobbles. This was our home for seven weeks of perpetual daylight. It sheltered us during periodic blizzards when we passed the time in our eiderdown sleeping bags reading *The Brothers Karamazov* or *War and Peace*. We never removed our undergarments, which were old-fashioned woollen combinations. It had taken nearly two weeks to get to Hinlopen Strait, all the way up the coast of Norway, then across the raging Maelstrom to Bear Island in a decommissioned sealer, and on to Svalbard, where we changed to a small boat called *Salterella* that dodged the ice floes around the island to take us to our field area. There we were abandoned. The World Service of the BBC was our only link to the outside world; there were not even radio communications with other members of the expedition. Health and safety regulations would not allow such isolation in the twenty-first century. 'All You Need is Love' was Top of the Pops every week while we were encamped at the Top of the World. With a twiddle of a knob we could tune in to Radio Moscow, which promised the end of capitalism and the triumph of Marxist–Leninist ideology, but the English hit parade was what we waited for. White Arctic foxes snuffled around outside the tent for the leftovers from our meals, which were all boiled up from dried ingredients. We drank lemonade made up from crystals dissolved in the pure water of the melt stream next to the tent. A few years later that stream would be christened Profilbekken; it had never had a name before. The sharp cries of terns and the sinister mewing of Arctic skuas accompanied the slow suck of the waves against the shore to lull us into sleep (strictly, at ten o'clock) even though the light hardly dimmed. Routine was an effective antidote to feeling alone in a vast and hostile wilderness.*

---

* I described the adventures of fieldwork in this remote area, long before the invention of the satellite telephone, in the opening chapter of *Life: An Unauthorised Biography*.

Geoff and I soon located the fossil-bearing rocks. They were well exposed along Profilbekken and extended along the shoreline towards the north. It was puzzling that they had not been discovered years ago. The landscape along Hinlopen Strait was dominated by the extensive stretch of raised beaches that flanked it landwards. This made for a particularly stony, bleak and uninviting habitat – nothing could live there. Even the polar bears seemed to eschew its windswept barrenness. Our theory was that geologists passing along the strait in their dories scanned the shore through binoculars and saw only this gravelly *strandflat* with the Valhallfonna ice sheet behind, and failed to notice the inconspicuous small rock cliffs at the sea's edge. They just shivered and passed on, until one day in 1966 when they needed fresh water. If an earlier geologist had made landfall he could scarcely have missed the fossils – they were everywhere! Loose pebbles on the beach showed fragments of trilobites, and once the bedrock was discovered a few hammer blows revealed the fossils in their rocky home. Hundreds of millions of years ago another, very different sea had been thronging with a variety of these marine arthropods, all long extinct, whose entombed carapaces now awaited discovery on one of the most remote beaches on earth. We were opening a window on to a hitherto unknown vista of the deep past.

Our camp was pitched near the youngest rocks in a thick succession of rock beds – hundreds of metres of strata awaited exploration. The rocks were tilted in such a way that progressively older strata lay northwards along Hinlopen Strait. We soon realised that those first, chance finds were just one glimpse of a whole sequence of diverse Ordovician marine worlds: collecting the full story was going to be a long job, and one that needed to be approached systematically. This meant locating collections precisely, and in their proper order, measuring rock sections, and then sitting down to extract the trilobites from where they had been hidden from view for so long. Occasionally, the rock came away in slabs with a whole trilobite served up to view like a kipper on a plate: ours were the first eyes to gaze upon a species 'new to science'. That was gratifying enough to distract us from our freezing toes, and celebrate with a smoke of the Wills cigarettes that were provided free to the expedition. More often, we had to smash tough limestone beds with our geological hammers to find fragmentary remains that could be dug out only much later in the laboratory. Few trilobites were as complete as the Moroccan one that began this chapter. Their carapaces fell to pieces when they died, and usually we uncovered just these fragments, particularly comprising heads and 'tails'. A whole 'dead' trilobite was something we longed for while we waited for an icy squall to pass. Some of our trilobites were so strange that we felt compelled to coin a field name for them – 'Fred' was a common find in dark limestone, sporting a headshield with a brim something like a homburg hat. We had no idea what sort of trilobite 'Fred' was, but we knew he was something special. It did not take long for us to recognise that there were several different species of 'Fred' with broad and narrow brims. As collecting went on we realised that we must have discovered dozens of different species of trilobite. It was a fossil bonanza. We were revealing whole faunas changing through time. Few palaeontologists have the privilege of discovering something so rich, so important and so novel. Nobody had dreamed that Spitsbergen would hide one of the most diverse histories of ancient life in the world.

It got even better. On some of the thinner rock slabs – often with one of the 'Freds' – were the unmistakable fossils of the extinct colonial animals known as graptolites. There was something in Spitsbergen for both the current and previous Woodwardian Professors: Bulman would have been as surprised by the graptolites as Whittington by the trilobites. Graptolites are possibly less immediately appealing fossils than trilobites, but they are very useful for dating sedimentary rocks with great precision. Their colonies were composed of many individuals occupying small 'tubes' that were added one after another to make remarkably symmetrical structures. Bulman and his contemporaries had proved to everyone's satisfaction that these colonies were part of the ancient plankton – which is why some species had such a wide distribution as they drifted over vanished oceans. They were preserved as fossils when they sank to the sea floor, and then the little 'tubes' occupied in life by tiny filter-feeding creatures showed up as a kind of fine saw edge preserved along each branch of the colony. Graptolites were rarely found in the same rocks as trilobites – their preservation often required conditions low in oxygen inimical to other life on the ancient sea floor. But Geoff and I found trilobites and graptolites together in our rock section along Hinlopen Strait, some of them shaped like crosses and a few inches in diameter, others resembling tuning forks, others again leaf-like – a whole gallery of different forms. They were easily collected from the thinner-bedded rocks on which they lay conveniently flat; we even knew a few of their Latin names. But if we didn't recognise exactly what our hammers revealed we did know that the graptolites made the work even more scientifically important. Our discoveries now came with an accurate chronometer. The Hinlopen Strait story would make a great doctoral thesis.

Just over halfway through the field trip everything changed. We had one visit from the expedition's small supply boat, partly to make sure we were surviving without mishap, but also to bring us our one mail delivery. The mail included the results of the end-of-year examinations. Geoff Vallance had a lower second as his final degree; I had a first-class grade in my Part 1B (my second-year assessment). The implications were obvious. I would be the likely candidate to take the discoveries we had made to the next level, in Cambridge, while Geoff would probably go elsewhere to earn a living. The field assistant would become the researcher. It was a profound disappointment for Geoff: the dice had spun in my favour. We were not natural friends, but in a small tent in the Arctic we were obliged to rub along as best we could. The excitement of discovery kept any major differences at bay (although we had argued about my taking an extra biscuit on one occasion) and the physical exertion of whacking rocks or marching through slushy snow all day did concentrate the mind on the essential business of how to process our dried food at supper into something palatable. I do not wonder that Geoff consumed much of his whisky ration after the boat had departed and chased me up the beach waving a piece of driftwood. Fortunately, he was too intoxicated to inflict much damage. To his credit, he went back to the rocks the next day and hammered out more trilobites. In a few weeks it was time for us to go back to the Norwegian town of Longyearben on Spitsbergen en route for the return to the mainland, and, ultimately, Cambridge. By then, the Arctic summer had given way to a perpetual gloomy pall, as if the sun had lost all enthusiasm for warming the North Pole. It glowered red, low in the sky. Our last task was to make sure our finds were safely packed in newspaper, and then stowed in collecting bags marked by locality, for passage to the Sedgwick Museum where the bags would eventually be unpacked for further study of the collections. After the 1967 expedition I never saw Geoff again.

*　　*　　*

Now at last I was fully focused. A trilobite-shaped deity was in charge of my destiny. My third undergraduate year was motivated by determination to get a final degree good enough to make those exciting Spitsbergen discoveries my future. I bade farewell to history and philosophy, although I was to discover much later that the discursive reasoning I had practised would become more useful than anything else when I began seriously to write for a wider public. Poetry ceased to be a preoccupation. My first-class passage through the examinations made me a senior scholar for my third year, and posh accommodation was one of its benefits. At last I had fine rooms close to the River Cam. My second year had been spent in ancient, but tiny rooms tucked behind the Arts Theatre, the bonus being that elsewhere in the same building I could hear six choral scholars rehearsing their way to becoming the perfectly harmonious King's Singers. Now, from the window of my study I could see punts elegantly idling along the river, and beyond it a rich meadow where cows grazed peacefully along the 'backs'. Nothing much had changed for a century; this was the Cambridge so often portrayed in period dramas. I discovered that famous people had lived in the same part of the college. John Maynard Keynes had roomed there before he became the most famous economist in the world; he returned to King's as bursar, so the money was in the right hands. E. M. Forster's initials were reputedly carved into one of the doors. My name was neatly painted in white letters on a black background at the bottom of the staircase: how many names of people more remarkable than me were hidden beneath it?

I met Bridget at a party in Selwyn College. The usual cheap white wine was steeping in a bath with ice cubes, and little bits of cheese and pineapple were impaled on cocktail sticks arranged on plates. The Supremes blasted out 'Baby Love' from a Dansette record player. It was too loud and too hot, but I didn't care. Bridget had come to the shindig with a good friend of mine, but she left with me; I believe I was overwhelmed. She had dark hair, lustrous eyes and a laugh I couldn't resist. I could not quite believe that one so beautiful could be interested in me. I was smitten, and there was no turning back.

Many of my college friends were left-leaning students reading English, who continued on to left-leaning jobs. David Leigh became a renowned investigative journalist on the *Guardian*. It is difficult to connect this laconic student working all night, cigarette in mouth, to deliver an essay at the last conceivable possible moment, with the fearless exposé journalist who revealed Jonathan Aitken as a liar. He hid his courage well. Simon Hoggart, the son of Richard Hoggart who wrote *The Uses of Literacy*, became a much-loved political sketch writer also employed on the *Guardian*. Simon and I frequented an upstairs tearoom on King's Parade, the Copper Kettle, where we pretended to be obscure members of the clergy. We christened our ancient waiter 'Sideboard'; he wore a well-rubbed plum-coloured dinner jacket and a bow tie. 'Ah!' I might say in an exaggeratedly sibilant vicar's voice, 'I wonder what sweetmeats the excellent Sideboard has for us today?' Simon would reply: 'After Canon Teabag-Montefiore's percipient sermon I confess to being a trifle overcome. I think I might relish a Viennese whirl!' 'Sideboard' became complicit in our game, winking occasionally at these silly young parodists. We were an amusing diversion for him in a dull job, and he rewarded us by bringing endless supplies of hot water to top up our teapot. We could keep our entertainment going for an hour, all for the cost of a cake and a pot of tea. We were sometimes joined by our Welsh Marxist friend, Gwynn Pritchard, who became one of the founders of the Welsh channel of the BBC. My engineering friends went on to be engineers, my medic friends went on to be doctors, and those who read law became lawyers – useful lives well lived. Scientists rarely joined the Footlights, so I never knew comedians who would go on to star on the television. The student newspaper *Varsity* regarded scientists as rather dull – 'grey men' – but in the long run they probably contributed well to society. Salman Rushdie became the most renowned of my contemporaries but I have to admit that the fact he was from Rugby School seemed much more important to me than his connection to India or his literary brilliance. I would know better now.

My flirtation with undergraduate politics was brief. As part of the leftward push in King's College there was pressure on the dons to have student representation on the College Council. Democracy demanded that the eternal traditions of the college should be scrutinised by the people! The groovy young economics don Bob Rowthorn was supportive: he was the very model of the new academic: clever, radical, handsome and with many girl-friends.* He was not afraid to stir things up with the older fellows. I think I might have been deemed the acceptable face of radicalism, because I found myself the first student representative. I was rather naïve about what that meant, and earnestly collected small grievances about such things as washing facilities or bar bills, concerning which I received responses from the college. My circulars addressed to my fellow students were mostly ignored, just crumpled

---

* As I write he is Emeritus Professor of Economics at Cambridge, and a Life Fellow of King's. *Tempora mutantur, nos et mutamur in illis.*

up by the pigeonholes in the post room. It wasn't long before I was removed by a palace putsch, and replaced by someone much more Marxist. My radical friend Gwynn Pritchard mumbled something about Kerensky and laughed, and indeed I had been removed by a classic revolutionary sleight of hand. I left politics alone after that, except to join the Grosvenor Square demonstrations against the Vietnam War in March 1968. I got quite bruised during a chaotic charge by the crowd, and had my name taken by the police. I am probably on file as an agent provocateur on some register buried deeply within MI5.

Colleges have grand feasts a couple of times a year, and King's was no exception. The Founder's Feast was the grandest of them all, when the wine steward went to the cellars to determine which claret was up to the task, and whether the '49 was a vintage port worthy of completing the performance. Candles illuminated the Great Hall; the portraits of past provosts looked down with appropriate hauteur from the walls. College silver twinkled extravagantly in the candlelight, its baroque excess somehow appropriate for the occasion. Cutlery was laid out in a complex hierarchy designed to confuse the lowly born. As a senior scholar I was expected to sit on the high table with the bigwigs. The provost of the college sat at the centre of the table looking out upon the longer lines of guests below. Edmund Leach was an anthropologist well qualified to cope with the quirk of human behaviour that demanded such bizarre rituals. Distinguished guests were invited, some of whom had to say a few words. In a nod to inclusiveness undergraduates were placed between the guests, and I found myself next to the celebrated Oxford philosopher A. J. Ayer. He was familiar to me from that television programme we had watched in Ealing when we were children: he was one of the famous intellectuals on *The Brains Trust*. Recently, I had read his work as the leading logical positivist during my stint as a philosopher of science, and I had admired his forensic acuity. In truth, I was overawed. On my other side sat the college wine steward, Kendal Dixon – a nice old gentleman, but profoundly deaf. 'What do you make of the burgundy?' he enquired. 'Very fruity,' my reply. 'Not as snooty as the '54 by a long chalk!' followed a knowing chuckle. It may have been nerves or plain greed that made me tuck into the burgundy with such dedication. As a selection of delicious wines accompanied successive courses I became more and more inebriated. Worse, a kind of combative truculence rose within me. Garbled recollections of my forays into Zen Buddhism elbowed out any common sense that might have held me in check. What was all this logical positivism stuff? There were truths out there that could only be accessed by letting go, by accessing the deeper reality that came with enlightenment: 'the sound of one hand clapping' and all that. I challenged the famous philosopher with my metaphysical insights. Reason only got you part of the way, or so I informed the ultimate reasoner. "S'all transhen-dental,' I lectured A. J. Ayer, taking my authority from the Beatles (who were in their mystic phase) and waving my hands vaguely in the direction of the blurry faces out there in the Hall. 'The times they are a-changin',' I added, unnec-essarily. I do not recall the philosopher's response, but I believe it was kindly. When a headache jarred me into consciousness the following morning I remembered enough of my foolishness to be mortified, and I am mortified even now. I have awoken many times in the night under cover of darkness and flushed hot with embarrassment at the recollection of my oafish gaffe at the Founder's Feast. Since 1968, I have never abandoned my adherence to rationalism.

*   *   *

I did get the first-class degree that was necessary to win a research grant. 'Firsts' were awarded sparingly, and I was convinced that a fellow geology student, Peter Alexander Marrack, was going to get one, and if there were only one it would surely not be mine. Peter did work extraordinarily hard and thoroughly deserved recognition for his diligence, and he was a nice person to boot, but I did manage to work up a kind of competitive lather in his presence which I understand is also true of sportsmen who feed on an antagonism to their opponents that lasts only as long as the match. In the end, *both* of us got the top grade. Professor Harry Whittington would now take me on as his research student. I may have been encouraged by the euphoria that followed my success, because Bridget and I agreed to get married by her father at his church in Arbury as soon as we could. The King's College senior tutor agreed that marriage was a good thing – a stabilising influence during the research years, in his opinion. Bridget's family – the Thomases – became my family, and her father Degwel inevitably filled the place of my own father, or rather showed beguiling Welsh warmth that I had never experienced before. Before he got the calling he had trained as a metallurgist, and Bridget was happy to bring a young scientist into the family. She had three older sisters, all with partners, and all on the political left. I felt comfortable with them. However, the immediate challenge was how to survive financially through the summer break until October, when the grant for my research scholarship was due. I needed a job.

It was rumoured that good money was to be made as an ice-cream salesman. Ice-cream vans were sent around the countryside from a depot hidden on an anonymous trading estate on the edge of Cambridge. It was a large shed tucked behind a premises where used car tyres were supplied with retreads. Most of the sales vehicles were small vans with the 'Walls' logo emblazoned on both sides. Ice cream was dispensed through a small window in the side of the van. The stock was kept chilled on a block of dry ice resting in a metal-lined case, and it was in danger of melting on sweltering afternoons. The firm's sales territory was divided up into strictly defined beats for each seller. Full-time employees were given the choice locations – like King's Parade just outside my college, which was always thronging with tourists in summer. The regular vendors were tougher than the Carter's boys – if anyone tried to poach their patch the dispute ended in fisticuffs. Rookies were allotted the villages and towns of the Fens, desolate places like March and Littleport. Prime targets were the council estates that surrounded the older town centres; these were jerry-built stretches of small houses constructed for the labourers who harvested the sugar beet whose lush leaves covered the endless flat fields of fenland. Nothing could have presented a greater contrast to the comforts of my ancient university. The little van had a wind-up musical box that played the same five notes over and over (a perky 'dong dong ding-ding dong') to announce our arrival. I came to detest that tune. The ice cream and lollies were bought from the boss in the depot to sell on for a margin – there was no basic wage, so everything depended on sales. I had to flog the stock before it melted or I would be out of pocket. Worse, I had a rival. The enemy was Milky Way, who had a larger van and a more interesting tune. If I approached the poorer end of some small fenland town Milky Way would come steaming out of the council estate just as I was about to cruise the same streets, triumphantly playing his tinny symphony at full volume to demonstrate that he had already cleaned up. My subsequent 'dong dong ding-ding dong' had a melancholy edge to it, a despairing superfluity. By the time I had paid for my fuel and cigarettes there was almost nothing left over. I had worked sixty hours for a few quid. I came to believe Milky Way had a spy on each estate who would tell him when he could sneak in to have the market to himself. The final insult was a visit to a particularly rough estate near Littleport that had already been fully supplied with ice lollies and cheap cones by Milky Way. I parked up on a grassy corner and wound up my signature tune for the hundredth time. A burly man with tattoos on both arms came striding across the grass.

As I leaned forwards through the window he grabbed me by the throat. 'I just bought bleedin' ice creams for the kids from Milky Way and I don't want you coming along and givin' 'em ideas. So piss off!' I did – and that was the end of my career as an ice-cream salesman.

My degree could not have been less relevant to my employment at the Chivers jam factory. As well as sugar beet, the fens were famous for fruit and vegetables grown on the peaty soils that had been established by draining the waterlogged ground three centuries earlier. Chivers was a centre for canning and jam-making housed in a large site on the edge of Cambridge. I soon discovered that a number of familiar brands emerged from the same plant – Hartley's and Rose's as well as Chivers all spewed out from production lines cranking away inside ranks of anonymous sheds. They needed labour, not degrees, but they did pay a wage. Employees had to clock in and out on a contraption that may have begun life when Charles Dickens was working at the blacking factory. I found myself on the new potatoes. My job was to pick off rotten or misshapen spuds from a never-ending conveyor belt before they went off to be peeled and cooked ('Hartley's New Potatoes with a hint of mint'). It was tedious work, and a challenge to believe that so many potatoes could exist in the known universe. At the end of the day I earned extra overtime by sweeping up the rotten potatoes and hurling them into containers for pigswill. Nothing was wasted, even if the smell was appalling. Near the end of the process mint flavouring was added to a huge vat of cooking potatoes from a flask of super-concentrated liquor. I never owned up to once accidentally dropping the whole bottle into the vat. My clumsiness would have had housewives opening tins of explosively minty spuds all across Britain. Perhaps they liked them that way. One day the potatoes changed to strawberries bound for Chivers' (or was it Hartley's? Or more likely both) strawberry jam. The conveyor belt was just as endless, although bad strawberries did not smell as nasty as rotten spuds. I also had the job of strawberry tipper at the front of the line, which was only slightly less relentless than that of sorter: pick up full tray – tip into the washer – dispose of the empty tray – pick up next tray . . . I understood now how working on a production line could so diminish the versatile organism that is a human being.

The women who worked in the factory wore unflattering floral over-garments that buttoned down the front. Hair of both sexes was enclosed in nets, but the managerial class wore white trilby hats and walked around with a superior air. One of the women on my line pointed out her colleague who was leering through an open doorway that gave into the next production line. That was Lil, she informed me, and Lil fancied me. 'She calls you Toffee Crisp,' she explained, 'because you look like the man in the Toffee Crisp advertisement.' I was uncertain what to do with this information. Fortunately, I was soon moved to the far end of the potato tinning process, where the tins rolled down from the canning machine still warm to have labels slapped on them, then to be packed into cardboard boxes by the dozen. The packing was probably more tedious than any of the other jobs on the production line. The tins, like the

Mississippi River, kept on rollin', kept on rollin' along. They rolled on their sides into the labelling machine that slapped 'Hartley's New Potatoes' on to each tin before rolling off to the packer. To relieve the tedium loud music was provided, one hour on, one hour off. I never thought I would *long* for Engelbert Humperdinck to sing 'The Last Waltz' but I found myself desperately looking forward to the end of the silent hour to relish the rich tones of Engelbert to get me through the next sixty minutes. I did not dare look up towards the adjacent shed lest I caught Lil's eye. I could sense her gaze in the back of my neck. I had never even eaten a Toffee Crisp. Somebody thought of a method of sabotage to introduce an extra break in the implacable tin rolling. The tins had to meet the labelling machine with their long sides transverse to the conveyor belt. If a tin was quickly turned lengthwise it jammed the labeller, which then endlessly slapped more and more labels on to the same lucky tin. The belt stopped, the alarm rang, and the whole system had to be reset. Five extra glorious minutes of repose, our own last waltz, stopping the Mississippi.

9

Getting Serious

The life of a British* doctoral student is almost ideal.
For three years the lucky candidate can immerse them-
self completely into the object of study with no
examinations to pass, no administrative responsibilities, no
distractions, except for chosen ones. The grant money is
just about enough to live on with nothing left over for
luxuries, but then three years of time is almost the ultimate
luxury. This is the period when the importance of the
college decreases as the importance of the department
grows. This is when the significance of past research is
appraised, criticised, and superseded. If being an undergraduate is often about summarising what is already known,
then the research student pushes out into new territory,
building upon, and sometimes upsetting the past.

I was allocated a research room on the top floor of the

---

* In the USA the PhD degree is almost invariably preceded by a Masters,
during which time intensive teaching continues.

Sedgwick Museum in Downing Street. It was reached by a lift of astonishing antiquity in which a stool was stored like a holy relic; it had been used by the famous graptolite expert (and pioneer female scientist) Gertrude Elles in her old age. Nobody dared to move it in case she came back from the dead. My attic room was shared with a fellow student, John Bursnall, who was working on the Ordovician geology of Newfoundland. Our room was lined with cabinets in which to house our collections and field specimens, so free space was very limited, and our desks were crammed into the interstices. The first job for me was to unpack the specimens that Geoff Vallance and I had collected on Hinlopen Strait. More than a year had passed since our adventures. The newspaper headlines on the wrapping paper already seemed historical: '"All You Need is Love" – New Beatles Single'. Occasionally, an outstanding specimen would be unwrapped, bringing back the moment of its discovery on a remote Arctic shore while we sheltered from the merciless wind.

Harry Whittington was my supervisor for my thesis on the Ordovician trilobites of Spitsbergen. I prefer the American term 'adviser' for this role because that was exactly what Harry did over the next three years – he offered advice, but left me largely alone to follow my own path. Bridget and I were immediately invited to his home to meet his wife Dorothy, an ebullient American whom Harry addressed as 'dearie'. She was as loud as Harry was quiet, and bubbled over with goodwill towards his students. They lived in a good if undistinguished modern house in a respectable suburb of Cambridge, where it could be

In Spitsbergen.

imagined that his neighbours were solicitors or bank
managers. Their drawing room displayed surviving knick-
knacks from their early years in Burma, where Harry had
begun his teaching career under the most challenging of
wartime circumstances, eventually fleeing the invading
Japanese army and escaping with little more than a few
precious mementos. He subsequently arrived at Harvard
University (Cambridge, Mass.) where he established a
distinguished research career by publishing relentlessly on
palaeontology (and especially trilobites) through the 1950s
and early 1960s. After he arrived at the other Cambridge
he established a research programme on one of the most
famous fossil deposits in the world – the Burgess Shale
– where are preserved an array of soft-bodied fossils which

would overturn our understanding of early animal evolution. Twenty years later the Burgess Shale animals were made world famous by Stephen Jay Gould in his book *Wonderful Life*. Although Whittington would have welcomed me on board the Burgess Shale project, my heart was with the trilobites, and the gift that had fallen into my lap under the shadow of the Valhallfonna glacier.

After unwrapping hundreds of specimens I had to arrange them in the order of the strata we had collected in the field; this would be the basis of my catalogue of geological time. I separated the trilobites from the graptolites. When I was sure of my ground I would show the latter to Professor Bulman. Drawers were labelled so that I could find what I wanted speedily in future: so far, so routine – and not so different from arranging my boyhood collections. The real work began with extracting buried trilobites from their rocky hideouts. Most of them had to be freed from the limestone that embraced them by manual preparation, painstakingly chipping away the enclosing rock with a mechanical tool employed under a low-power binocular microscope. The extraction instrument works by vibrating a needle of strengthened steel that is harder than the rock, which then tends to flake away from the surface of the fossil. It is slow and laborious work, and easy for a beginner to make a mistake – one slip and the surface of a precious specimen can be horribly gouged. I started on a trilobite that was well known from several localities around the world. From the illustrations of other species I knew what I should find beneath the surface of my new specimens – I had a pattern to follow. I spent many weeks refining my skill in preparation. There was something therapeutic about it, because absolute concentration was required – if the mind wandered the specimen was likely ruined. This was my new Zen. I became skilled in extracting delicate spines without damaging them. I delighted in the subtle patterns that some trilobites showed on their surfaces: arrays of lines like fingerprints, terraces, tubercles, pits – it was a measure of my improvement that I could extract such fine features without damaging the rest of the specimen. John Bursnall called it 'sculpting my animals' but the poor man was driven half crazy by the perpetual buzzing of my modified engraving tool. I was having fun, and he was having to hide.

This was only the beginning. I still had to piece many of the trilobites together. Complete specimens were rare, but they were invaluable when they were found. Trilobites moulted as they grew, like most arthropods, and the majority of fossils were just their shed parts: heads came into four main pieces, the thorax fell into individual segments, leaving the tails behind. To get an idea of the whole animal the fragments had to be pieced together like a jigsaw puzzle. There were clues – for example, certain kinds of head came with a distinctive type of tail – but there were many puzzles that could be solved only by careful detective work. A special type of surface decoration might link two separate parts, or two pieces might articulate in only one unique fashion. Hundreds of specimens took months to work on. I mention these details to show that effort and subtle observations went into piecing together the different species of trilobites from the

Ordovician rocks of Spitsbergen. Eventually, well over a hundred different kinds of trilobites were discovered from those remote cliffs and streams. Many of them were new species. Only when they were identified could I think of giving them names.

Naming fossils is one of the minor pleasures of palae-ontology. All animals have a binomial scientific name, often Latin or Greek in origin, and they have to have an 'official' name before they are accepted as genuine species. Names in theses are not formally published and remain provisional until they are. It is not permitted to use the same name twice, so ingenuity is sometimes required to ensure orig-inality. Nearly all my PhD thesis names were eventually published. What we called 'Fred' in the field became two genera. 'Broad-brimmed Fred' became *Balnibarbi*, named for one of Lemuel Gulliver's travels to a land peopled by eccentric natural philosophers – this was Jonathan Swift's satirical take on the Royal Society (my favourite Balnibarbi mad experiment was an attempt to distil sunbeams from cucumbers). 'Narrow-brimmed Fred' became *Cloacaspis* from the Latin for sewer (*cloaca*) because its fossils were found in black limestone stinking of sulphur. The broad 'Freds' included one species with an exceptionally wide brim, which was called *Balnibarbi sombrero*, which requires no further explanation. *Oopsites* was derived from the Greek for 'egg eyes' for a big-eyed trilobite, but also reminded me of an occasion when I accidentally destroyed a good specimen and 'Oops!' seemed appropriate. *Gog* was a giant trilobite named after a giant. There is a lot of fun with names, but a serious purpose, too; once they are given to well-defined species they become 'data'. Changes in the number and distribution of species can be plotted. Counts of individuals can be made to discover if a few species are dominant. The main point is that these analyses are only as good as the data (names) that go into them. The months of hard graft going into digging out specimens is just as 'scientific' as computer programming or setting up experiments in a chemistry laboratory. It is the dogged work that validates or disproves any theories that grow out of the results.

Once I had made decisions about species and named them, I began to plot out the trilobites against the rock strata we had measured, and patterns began to emerge. There were synchronised changes in the trilobites through time – not all of them were found together in one place in the rock section. Instead certain trilobites were found associated in sets of particular rock beds, while other species chummed up in different parts of the section. 'Freds' (all of them) were found with other members of the same trilobite family (Olenidae) in two thicknesses of strata separated by a very different assemblage of a dozen or so trilobite species that dominated through at least thirty metres of strata, representing a few million years of geological time. There had to be an ancient environmental control over their preferences. Trilobites came in 'packages'. Some species preferred to live together, and avoided another marine environment that well suited 'Fred' and the associated Olenidae. What kind of habitat control could this be? It was obvious too that the 'Fred' strata were also those that yielded the most abundant graptolite fossils.

These rocks were all very dark – even black when fresh – and many samples emitted a sulphurous smell when they were split. It left me wondering what strange sort of world these ancient animals inhabited.

*　　*　　*

A revolution in scientific understanding was happening in Cambridge during my undergraduate years that would ultimately help to explain many details of my own research: the belated acceptance that the continents had once been joined together in the 'supercontinent' – Pangaea – that then split up as its components 'drifted' apart to reach their present geographic configuration. Although Alfred Wegener promulgated this idea in the 1920s it had not achieved wide acceptance until the ten years before my arrival in Cambridge. Edward Bullard had prepared remarkable 'best fit' maps showing how closely the profiles of the edges of the continental shelves on either side of the Atlantic Ocean fitted together like the pieces of a jigsaw puzzle. This result was achieved with the help of computers, which seems routine now but was then at the cutting edge. Bullard worked at the geophysics laboratories up the Madingley Road, quite removed from the quiet world of the Sedgwick Museum. When the mid-ocean ridges were mapped, it was realised that this was where new oceanic crust appeared as the continents moved apart ('filling the gap'). The analogy that this happens at the same rate as fingernails grow has been repeated so often that it is becoming a cliché, but it does carry the right note of inevitability. What had originally looked like speculation began to look like an undeniable fact. The earth's surface could be regarded as an assembly of plates that moved relative to one another, drawing apart in some areas of the world, while colliding in others to throw up mountain ranges as high as the Himalayas; an inexorable earth motor. This view of global geological creation and destruction became dubbed 'plate tectonics' and soon entered the language. It has been adopted as a metaphor by opinion formers in newspapers who wish to convey something that moves with inevitable consequences, albeit very slowly, but in the 1960s it still had the lustre of a freshly minted idea. Plate tectonics seemed to explain so many things at a stroke: why volcanoes were where they were; how and where linear mountain chains formed and why they spawned granites; why similar fossils were found in Africa and South America, so far apart today. It was a theory of almost everything.

Dan McKenzie was one of the crucial scientists in generating our understanding of the geometry and movement of tectonic plates. Dan was a geophysicist a few years ahead of me in King's College. During my final undergraduate year I was always vaguely aware that this bright star glowed sufficiently to put others in the shade, and that he was the first to apply Euler geometry to describe the movement of the plates. He was already becoming famous when his PhD thesis was hot off the typewriter. Extra copies had to be made to accommodate the demand. Nowadays, Pangaea and its break-up is just an interesting fact accepted almost without question. I was at Cambridge at exactly the time when the complexities of plate tectonics and the fate of continents was still hot stuff. As for Dan McKenzie, he has been there all my life, a tall and diffident figure out in front, always getting there first. One of the early lessons of the scientific life is that there will inevitably be somebody cleverer than you receiving invitations to speak at Harvard University while you are struggling with the final chapter of your thesis. It hurts a little at first, until you realise that science needs officers as well as generals. If it is not stretching the metaphor too far, science is a war on ignorance and a campaign requires all ranks for a successful outcome. The legions follow the generals with their own discoveries, each small victory confirming the justice of the cause. When the war is won, the general gets the laurel crown and the troops share the buzz to know that they were on the right side. Science moves forwards, not caring one jot about the warriors, whether generals or foot soldiers.

Shortly before I embarked on my seminal trip to Svalbard a short scientific paper by J. T. Wilson was published in *Nature*. It changed the way I interpreted my results. Simply entitled 'Did the Atlantic close and then reopen?' Wilson's little paper stated what now seems to be obvious. Plate tectonics did not begin with Pangaea and its break-up. There was a more distant period of earth's history when Pangaea itself was assembled from even older continents that were stitched together by the same tectonic forces that move the plates today. The face of the planet was in continuous motion, at a pace both slow and implacable. The dance of the continents was staged over hundreds of millions of years; the geography at the period of deposition of my Ordovician rocks must have been utterly different from that when mighty Pangaea was assembled nearly 200 million years later. The Ordovician was a time when the continents were spread as widely across the globe as they are today – although then they were *different* continents ('palaeocontinents' to give them their scientific label). Old conundrums were plausibly explained: the western Highlands of Scotland had always looked geologically like a piece of Canada and New York State. This chunk of Scotland was stranded on the European side of the Atlantic Ocean when it opened in the era after Pangaea. Back in the Ordovician it was a part of the ancient palaeocontinent of Laurentia and located in the tropics. A trilobite could have swum from Albany, NY to the Isle of Skye through a warm and shallow sea. The research programme switched to reconstructing the continents as they were before Pangaea was assembled. New jigsaw puzzles for solving older worlds.

In the office next door to mine in the Sedgwick Museum an exceptionally talented, energetic young lecturer was exploiting the new understanding of plate tectonics to account for all manner of features of ancient mountain chains. The Appalachians continued from North America into Scotland and Norway and were once a single entity. When Pangaea was intact this mountain chain snaked through the supercontinent rather in the way the Urals cross Russia today. This mighty ancient mountain chain had been created when ancient continents collided – analogous to the way in which the Himalayan chain is the consequence of the push of the Indian subcontinent against the vast body of Asia. John Dewey and his American friend Jack Bird were milking plate tectonics with gusto to explain all manner of features along what is perhaps best called the Appalachian–Caledonian mountain chain. The two young scientists cackled and joked together in uproarious fashion, while I buzzed on my preparation tool to extract my trilobites. Poor John Bursnall must have been driven to distraction by his noisy neighbours. Dewey was slightly podgy then, but unusually athletic. When particularly excited (which was often) he might suddenly flip into a very professional handstand, which he could hold for a long time. He spoke as fast as he thought, which was very fast indeed. He had an apparently endless capacity to absorb information, build it into the big geological picture and then prepare explanatory drawings of elegance and lucidity to publish the results. The process of drawing helped to clarify his ideas in a curious kind of feedback loop. In my own work, too, drawings were becoming an important part of my thesis; those afternoons in the Art Room with Mr Bland were proving their worth. Dewey loved fieldwork,* which he regarded as the basis of good geology; he liked to quote the pioneer structural geologist E. B. Bailey: 'The best geologist is the one who has seen the most rocks . . .' So here we have three scientific

---

* John Dewey had been concerned that young geologists were relying on instruments and theoretical modelling rather than field observations. He established a medal at the Geological Society of London to recognise outstanding achievements in field-based work.

sketches: McKenzie, mathematical and theoretical, but aware of how his ideas might be tested; Dewey, arch synthesiser of observations made in the field by himself and others; and myself, at last seeing how my own fossil work might mesh with the change in the zeitgeist.

Harry Whittington appeared from time to time during this creative phase. He tolerated my mistakes in deference to my enthusiasm, occasionally nudging me towards an obscure trilobite reference I had missed. Some 'advisers' regard their research students as little more than satellites in their own, substantial planetary system, but Harry was content that his students were making their own orbits. He did occasionally indicate his doubts about some idea or other by rubbing one side of his chin and throwing a dubious glance – it was enough to encourage second thoughts. Like all his former charges, I think of him with affection and gratitude. Some of his colleagues did not share this opinion. They mistook his gentleness for feebleness. They even tried to oppose his appointment. One of this number was Norman Hughes, a specialist in fossil spores and pollen, who clearly loathed the Woodwardian Professor. Hughes had a generously proportioned bald pate and a fleshy mouth and thought of himself as humorous, and indeed his whole body shook rather frequently with laughter, but never at his own expense. When I talked with him over morning coffee among the old building-stone collection on the ground floor of the Sedgwick Museum, I foolishly began a sentence with 'Professor Whittington thinks . . .' Norman Hughes cut me off sharply. 'Jellies don't think!' he snapped, immediately segueing into one of his laughs; a Cambridge put-down at its most vicious. Harry Whittington went on to win the Japan Prize many years later, probably the nearest thing to a Nobel Prize in palaeontology. Few people remember his detractor, other than by the soubriquet 'Norman the spore-man'.

Thesis research just got more interesting. When I tested the reaction of the dark Svalbard limestone in acetic acid the calcium carbonate dissolved, and the rock slowly disappeared leaving behind only things that were insoluble in the weak acid. From one of my samples, graptolites floated out. My heart leapt. The usual preservation of graptolites is flattened on the surfaces of shales. Colonies of these extinct, floating animals came to rest on the sea floor, where they were covered by sediment and squashed. Their original organic walls rapidly changed to a carbon film: they became more like shadow puppets of the original animals. Living graptolites had colony walls made of a collagen-like material that was delicate, but also insoluble in weak acid. This original material had survived in my samples. When the limestone that had embraced them was dissolved away they floated free, almost as they were when they were alive.* In this preservation they showed extraordinary details all but invisible on the usual flattened specimens. The previous Woodwardian Professor, O. M. B.

---

* What was revealed was the arrangement of tiny tubes in which the original plankton-feeding animals were housed. Since the animals (zooids) themselves were entirely soft-bodied they are never seen as fossils. Isolated graptolites also include their small growth stages, providing unparalleled clues to the development of colonies from one founding individual.

Bulman, had made his name working on similar material that he had extracted from younger Ordovician limestones in Scotland. So now I had something of great interest to two world authorities working in the Sedgwick Museum, Cambridge – graptolites as well as trilobites.

I soon discovered a graptolite that had never been seen before. Many Ordovician graptolites had branched colonies, with the little tubes occupied by the living animals arranged in lines along each branch – hence the resemblance to hacksaw blades when they were fossilised sideways-on in the usual way. In life, the colonies filtered out tiny plankton for food as they drifted far above the Ordovician sea floor. Some graptolites developed a different arrangement, such that the lines of little tubes were arranged in series back to back, looking somewhat like the grain-bearing spikes of wheat or barley. Two different configurations of these so-called scandent graptolites were familiar, with either two or four series of tubes in contact (termed biserial or quadriserial, respectively). What had never been seen among dozens of species was any graptolite with *three* series back to back (I would have to term it 'triserial') – but this is exactly what I thought I had discovered. At first, I had only fragmentary material. Somewhat nervously, I approached Emeritus Professor Bulman's office, which was as any good palaeontologist's should be, crowded with specimens and books. It was surprising the whole place had not gone up in flames because Bulman was a tremendous smoker, and splashed alcohol freely over rock slabs covered in graptolites to see them more clearly. The fumes were everywhere. He gruffly acknowledged my presence as I mumbled something about triserial graptolites. I handed him my finds and he coughed a little while he examined the fragments under a low-powered binocular microscope. 'Not convinced,' he said in a way that implied that further conversation was not required, 'triserial graptolites don't exist.' I retreated wounded, but not defeated. Over the next three weeks I dissolved many more blocks of the right kind of limestone in acetic acid. At last, I was rewarded with the perfect specimen, which floated out from its rock prison intact – a complete colony several centimetres in length and indubitably triserial. I gingerly removed it from the neutralised acid bath into a phial of glycerin for safe keeping. I might make a more modest assessment of my discovery today, but when I found my graptolite I could not have been more delighted if I had found the common ancestor of humans and great apes. If it had been a dinosaur as novel, it would have made newspaper headlines. This time I was sure of my ground, and Professor Bulman was astonished – he said something like 'Well, I never!' and his manner changed immediately. 'This must be written up.' It was an order. That is why the subject of my first solo scientific paper is a graptolite, rather than a trilobite. It may not have made the newspapers, but it made my life worthwhile.

Bulman set me up in a laboratory in the basement of the Sedgwick Museum. Drawing isolated graptolites required the utmost stability, and the old attic of the museum had ancient floorboards that moved enough to disturb the graptolites floating in their Petri dishes. A sketch

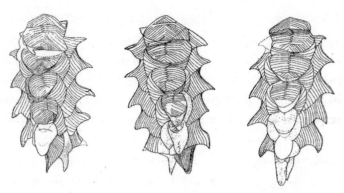

My drawings of the special graptolite from the Ordovician rocks of Spitsbergen from my first solo scientific paper.

made with a *camera lucida** required the subject to stay absolutely still, and the concrete basement was fit for the purpose. John Bursnall must have been delighted to get rid of me for a while – peace for him at last. For me, it was like being back in the Art Room at Ealing Grammar School for Boys. I recovered some of my skills: my 'object' in this chapter is one of those drawings. I was completely engrossed in the task for some weeks, all of that time being a digression from my PhD. I then had to write the scientific paper on the first triserial graptolite.

As a natural mimic I soon learned the 'voice' of the scientific journal. The personal must be excised completely – nothing about how thrilling it was to make a new discovery. 'I' becomes 'the author'. The introductory part of the paper briefly explains the history of the topic and

---

* A *camera lucida* is an optical device on a microscope that allows the viewer to see the object under examination and pencil and paper simultaneously, helping accurate drawings to be made.

outlines the importance of what is to follow, with references to the work of predecessors – let us say Snooks (1934). The full bibliography is at the back of the paper, so that is where you look to find out where Snooks published his findings. You are absolutely not permitted to say:'This paper proves that Snooks (1934) got it entirely wrong (the fool)' but you *are* allowed to say: 'Snooks (1934) interpreted the structure of the graptolite as biserial whereas the current work shows that it is significantly different . . .' This formal reticence does avoid confrontation in an insulting way – but of course that does not prevent a criticised scientist from being affronted. 'Materials and methods' usually follow in a palaeontology paper, and after that the meat of the work with new information and observations, illustrated with care. If a species is being named or described there is a formal way of laying that out, too. The material on which a species is based should be listed and curated in a recognised, permanent collection; if a new species is proposed it must be distinguished from all existing ones, and its name must not be insulting. *Snooksia incompetens* would not be acceptable. The paper usually finishes with some sort of discussion explaining the significance of the work, and subtly giving old Snooks another kicking. Conclusions end the text, except for acknowledgements, which tactfully express gratitude to the professor for his sage advice. At the top of the whole thing an abstract summarises the main points for those who want to know the gist without having to read the details. These days, a paper is usually published online before it appears on paper (a paperless paper is no longer an oxymoron). Date of publication is important in 'hot' science like particle physics because it establishes priority. In a competitive world, the rewards go to the swift. There was nobody out there with a triserial graptolite – no one even suspected it existed – so I did not have to worry about being pipped to the post.

Professor Bulman read my draft and approved it. I received a brief but friendly smile when at last he took the cigarette out of his mouth and placed it smouldering atop a pile of readily inflammable offprints. Bulman set the bar high and I had cleared it: if I had ever entertained regrets about not becoming a historian of other people's science they disappeared at that moment. I owned a discovery. Only one hurdle remained, to submit the paper for publication in the journal *Palaeontology*. Thus began a routine that would be familiar to any scientist: a threshold that had to be crossed to signal a new seriousness. The paper was submitted to the editor. He or she sent it on to an independent scientific referee (generally, two referees) who chose to remain anonymous. Eventually, their comments and an evaluation were returned to the editor. Some papers were rejected at that stage, but otherwise the comments were sent back to the author for consideration (usually, they improve the manuscript). If the editor sees no problems the revised paper joins the queue for a number of the journal. When I started to publish this was often a leisurely phase, so my contribution to science took over a year to appear in print. Electronic publication has speeded up the whole business in the twenty-first century,

but the same procedures apply, usually described as 'peer review'. This remains the best guarantee of quality science.*
Harry Whittington was uncharacteristically insistent on the necessity for publication. 'If it isn't published, it does not exist' became something of a mantra for him. He observed this rule himself throughout his long life. The phrase 'publish or perish' was already current when I was a graduate student, and it has since come to dominate academic life. No longer can an unworldly don wave his briar pipe vaguely in the direction of a pile of handwritten notes and proclaim that the great work was fermenting nicely. Results are not allowed to mature in the bottle. Nonetheless, I was unusual at the start of the 1970s in having a published paper or two on my curriculum vitae before I had taken my doctorate.

This was the most single-minded time in my life. We lived within walking distance of the Sedgwick Museum. Every day I crossed the bridge over the River Cam and traversed Midsummer Common towards Downing Street, a pleasant amble through an attractive part of the old university town. The regular routine was a pleasure, the more so because memories of packing potatoes were still fresh. Bridget worked in a travel agency, and together we had just enough to get by. Quite soon, a baby was on its way. I do not believe I did anything concerning fungi or wildflowers during my thesis years. I certainly did not write anything that was not concerned with palaeontology. I was turning into a

---

* Referees, and many editors, are unpaid. The online era has spawned a mass of predatory journals that ignore peer review. They will publish almost anything – for a fee.

specialist. There were new friends, but not from King's College. Fellow research students and 'postdocs' built a different social circle. Young married couples hardly ever lived in college; indeed, another kind of Cambridge lay outside those ancient walls.All the research students I knew were from relatively ordinary backgrounds. Upon graduation the Etonians and their congeners had all left for the City or politics. The gossip among my peers was now of geology and the ruses needed to get a permanent job, something that was already starting to be a challenge. I could talk endless trilobite details with two new postdoctoral fellows who had come to work with Harry Whittington. My undergraduate friend Michael Welland returned from Harvard University (where he had earned a Masters degree) with Carol, his young American wife; he started researching a thesis on the geology of Greece. Carol had come from a superior East Coast ladies' college, Mount Holyoke, and found the introspective snobbery of Cambridge something of a shock; her friendship with Bridget saved her from bolting back to the USA. I was becoming just a little more grown up; but the breadth of vision belonging to the curious boy I had once been was sacrificed for an orderly fascination. I was too busy to notice my loss. I enjoyed the extended family that came with Bridget, that made such a contrast from the claustrophobia of suburban Ealing. It was regrettable that my young wife did not get along with her mother-in-law.

By now, my mother had developed a troubled restlessness that was part of an uneasy adaptation to her widowhood. She moved from Ham to the neighbouring village of Shalbourne into a modern house that was easier to run than ancient Forge Cottage with its deep thatch. She had acquired a kind-of companion, John Norton, who lived under the same roof for a number of years. I believe they met in the village pub where I had been hailed as a class warrior. He was a retired civil servant with a big drink problem. He did useful jobs around the house and garden, and willingly took on such chores in return for free accommodation. He probably recognised that being with Margaret saved him from perdition. He was broadly cultured and so was able to offer proper companionship on expeditions across Wiltshire and beyond. His status was curiously undefined, but my mother was certainly in charge. John followed her in her subsequent house moves, continuing to live in a grace-and-favour capacity until he died. A particular difficulty was that my mother did not approve of Bridget and Bridget did not approve of my mother. I was glad of John Norton's role in moderating some fraction of this enmity. My principal loyalties had been captured by my new, extended family in Cambridge. I avoided any confrontations at my former home, and rather than face up to conflicting emotions buried myself deeper and deeper into the Ordovician period. I now recognise that this was not so different from using art and music to deflect attention from my deficiencies as head boy of Ealing Grammar School: I had a talent for blotting out what I didn't want to see. I remembered how my father continued to cast flies for wild brown trout, all the while ignoring the tax inspector waving papers right in front of him. Such a strange variety of amnesia may have served me well in writing my books on scraps of paper in airport terminals or while carrots burned on the stove. It could, however, have unexpected consequences.

*   *   *

The evolution of the ancient continents in the era before the unification of Pangaea became the subject of everyday conversation over coffee in the Sedgwick Museum. There was something paradoxical about the setting. Coffee tables were set out in the centre of a unique collection of building stones arranged along the walls, almost to the ceiling. Nearly all of these samples were rectangular, polished slabs of rock identified by neat, but decidedly old-fashioned labels beneath: 'Shap granite', 'Cotham marble', 'Paludina Limestone (Purbeck)' and so on. It was the epitome of a nineteenth-century museum, regimented and generally rather dark, but the talk around John Dewey's table was all about island arcs in a 'proto-Atlantic' ocean, and Alan B. Smith was explaining his pioneering reconstructions of the earth as it was 450 million years ago using newfangled computer technology. If the old museum had a fusty atmosphere, fresh conceptual air was blowing in from all quarters. Research students wanted to find their place in the new scheme of things. My discovery of trilobites and graptolites may have been the basis of a growing doctorate, but there had to be connections with this new world order. Even the traditionally separate departments focusing on geology, mineralogy and petrology,* or geophysics were beginning to feel like a single entity bent upon the overarching purpose of unravelling the narrative of our planet. They would become united as a grander and more ambitious Department of Earth Sciences. Plate tectonics heralded this time for new syntheses. The history of life had to be intimately entwined with the history of the earth, and understanding more about the workings of tectonics would surely illuminate the profound questions being investigated by palaeontologists. Everything must be interconnected, driven by that deep internal motor that continually rearranged the continents and reconfigured the oceans.

I needed to find the place of my discoveries from Spitsbergen within this new consensus, and to understand how Ordovician geography influenced life in the oceans well before it colonised dry land. New insights are sometimes portrayed as akin to the enlightenment of St Paul on the road to Damascus, a blinding flash, perhaps, or, to use another biblical image, a time when scales fall from the eyes. I have been scouring my memory for such a numinous intervention and I have failed to find it. This is disappointing, as a eureka moment adds glamour and inevitability. Rather, I gradually tumbled to a satisfying explanation of patterns I had recognised over months of patient research, as I came to know more and more of the fossils that I was unearthing. After my graptolite diversion I had to get back to the

---

* For some years the professors in geology and 'min and pet' were at daggers drawn, and the door connecting the two departments was locked.

trilobites on which my thesis depended. Only when I had identified and named all the players could I appreciate how they were acting in the geological drama that I was slowly translating. I have mentioned that the trilobites from the strata along Hinlopen Strait fell into 'packages'. I soon real-ised that there were three different communities (as I later termed these 'packages') of trilobites, each one rich in species, and each adapted to a distinct habitat. I recognised that they could be placed in an order that made sense if they were arranged across a profile ranging from shallow to deep water. The shallow-water community never crossed into the deep-water community; but both intergraded with an intermediate, and particularly diverse community, containing a third set of distinctive species. The deep end was where the Olenidae (including both kinds of 'Freds') abounded, almost to the exclusion of anything else. There was evidence that the sea floor was quiet in this habitat: moulted growth stages of some species showed shed pieces of carapace that lay undisturbed by currents, and these were found alongside delicate graptolite colonies. It made sense that they lived, and then accumulated as fossils, well below the influence of waves. By contrast, energetic currents had disarranged the shallow-water community into a melange of trilobite fragments, rather as shells lie piled upon a beach. Some specimens were even broken. The intermediate community was more mixed but there were occasional whole trilobites to thrill the aficionado, although the strange black deeper-water graptolite limestone was absent. The Olenidae had almost disappeared, but a host of different and lovely trilobites replaced them, including the giant of the fauna I had christened *Gog*. I realised at some, but still slightly mysterious point that the profile I was describing could be explained by invoking what was being discovered about ancient Ordovician continents. Surely, at the edge of those continents there would be continental shelves, with shallow to deep sea-floor profiles marking the edge. This was a simple and convincing way to explain my onshore-to-offshore depth gradient. Different communities of trilobites liked to live on different parts of the shelves surrounding the ancient continents, a broad preference that ecologists still recognise in oceans today, although populated with an entirely different cast of biological characters. I do not recall shouting out 'Gee whizz! Hot dog!' as such a moment of enlightenment might have been branded in 1970s Hollywood, but I do recall a feeling of inevitable rightness. It was obvious when you thought about it. I realised how lucky (again) I was to have discovered a rock section in which all these communities were interlaced. In its own way, Hinlopen Strait was proving a geological Rosetta Stone.

I would have to explain the idea to Harry Whittington. This made me nervous, as he had published a review in 1966 in which he had failed to find any particular connection between trilobites and past marine habitats. I was going against the 'prof'. Harry was characteristically generous when he heard me out: my explanation made sense to him, too. By then, he had moved into the Burgess Shale project, and perhaps he was not so wedded to his previous work, but many supervisors would not have been so accommodating. Over the next decade or two I was able to explore the implications of my not-quite eureka insight in many different ways. I used the shallow-water trilobite communities to mark out the extent of Ordovician continents, almost as if they were postage stamps that could identify a former kingdom. I mapped the deeper-water communities to find the boundaries of those vanished continents, where special trilobites and grapto-lites lurked in rocks deposited only around their edges. I explored the habitat where the Olenidae thrived, and began to understand how some trilobites thrived in environments unusually low in oxygen. I investigated the few trilobites that ignored the sea floor altogether and became part of a planktonic community. This is not my story here,* but it needs to be said to show how fecund a single, even simple insight can become if its implications are followed through. None of these creative offshoots would have been conceiv-able before plate tectonics in the context of deep geological time. The scientific papers I wrote one after the other in the years that followed were responsible for giving my scientific career momentum just when I needed it.

What I needed first was a job, especially after our son Dominic was born early in 1970. A job duly appeared, presumably opened up by the same minor god that had made me a devotee of trilobites, sent me to Hinlopen Strait, and appointed Harry Whittington to Cambridge University.

---

* I explained many of these fascinating interactions between trilobites and plate tectonics in my 2000 book *Trilobite! Eyewitness to Evolution*. I was soon competing with other palaeontologists pursuing similar ideas: science always moves fast when there is a change in world view.

The job was at the British Museum (Natural History) – as the Natural History Museum in South Kensington was then officially called – behind the same polished doors I had first broached with my Jurassic ammonite from the Dorset cliffs even before my voice had broken. The job was to be the museum's 'trilobite man'. The position became vacant only because Bill Dean, the previous trilobite man, had left for employment in Canada when he was not promoted to 'keeper' – the museum's term for a departmental head. I had one disadvantage. The doctoral thesis still had almost a year to run before it would be complete. There were other applicants who had PhDs already under their belt. I knew them all, and in time they became good friends and colleagues. We made a jittery queue outside the Board Room, hidden deep inside the Gothic cathedral dedicated to wildlife in South Kensington, and we waited our turn to be grilled by a small number of men in suits around a vast polished table. In my case the grilling was not too unpleasant, although when I was asked about my sporting prowess by the man from the Civil Service Commission all I could offer was tiddlywinks, which earned general guffaws. I was offered the job, despite my youth, but I was appointed as a research fellow rather than immediately becoming an established civil servant. I guess they wanted to make sure I would be worth a salary in a few years' time. I was, to use a superlative of the time, distinctly chuffed. I have recently wondered whether Professor O. M. B. Bulman had anything to do with my appointment. He was a trustee of the museum at the time, and maybe he had had a word or so in an ear or two about the young fellow who found the triserial graptolite. Things like that happened in those days. I do know that I nearly suffered a seizure when I met Bulman in the museum lift after I had moved to London and he actually addressed me as 'Richard'.

So began my working life in London at the Natural History Museum, an association that has lasted for fifty years. When the government finally stopped paying me I was able to spill the beans about this unique place in a book, *Dry Store Room No. 1*, but to begin with I was simply awestruck by the size and complexity of the esoteric world behind the scenes. Away from the public galleries, the museum was a kind of maze, constructed around specimens: it was easy for a newcomer to get lost in corridors that went nowhere, or doors that opened on to yet other doors. Collections numbering millions of examples were housed in rank upon rank of wooden cabinets across five departments. Just to open, glance at, and close the drawers of neatly pinned butterflies in the Entomology Department would have taken several days. The Botany Department at the top of the museum was like a temple stuffed with uncountable numbers of herbarium sheets, each one a thing of beauty in its own right. There were rooms full of mammals and molluscs, midges and minerals; dinosaurs, dragonflies and daisies. In offices close to the collections scientists worked away to become the world expert on their chosen subjects. I was now responsible for the national collection of trilobites. I could hardly believe it. As I wrote in *Dry Store Room* it was as if somebody had told me: 'Amuse yourself. For money.'

I now had to commute by train into London, initially from Cambridge, which was a slow journey at that time. One stop down the line the same men got on every day at Audley End and sat in the same seats. One of them was always greeted with the line 'Morning Minister. How's the portfolio?' but I never found out who he was, or whether he really had a portfolio. Now I was one of the working stiffs, but I was the only one with a job like mine. I hugged the thought to myself, even as life was throwing up challenges elsewhere. Dominic had been born with a dislocated hip, and the socket was not properly formed; to correct the fault an operation of some complexity was obligatory. A good orthopaedic surgeon working in the Battle Hospital, Reading, provided a compelling reason to move to a small town in the Chiltern Hills, west of London – Goring-on-Thames. I still had to commute quite a distance by train to the Natural History Museum. Little Dominic had some ghastly plaster contraption to wear after his main surgery, which splayed his legs to look like those of a swimming frog. Long, ineffectual hours were spent in hospital. Elderly ladies would peer expectantly into his pram, and then say 'poor little mite' with genuine concern. However, I still had my PhD thesis to write, and write it I would, come what may. Every evening I sat at the desk in our rented flat in a nice old house and pounded out a minimum of 300 words on my faithful Brother Deluxe portable typewriter; a talent – if that is what it was – that tested to the limit my ability to block out a world I did not want to confront in favour of dogged production. Commute and job followed by commute and work did not leave enough time for family or moral support. I don't know whether to admire my steadfastness of purpose or deplore my inability to cope with the complexity of real life. Whatever seemed most difficult I attempted to ignore. I did deliver my thesis on time, and Dominic's operation was successful, so that any damage that may have been done during this period could be disguised within an aura of success, or at least considered a price worth paying.

Doctoral examinations are relatively modest affairs in Britain. In France and Belgium they are big events that are held in public, where family and relatives sit as witness to an unveiling of the thesis as all the relevant protagonists are lined up on stage. The candidate offers a 'defence' of the work dressed in his or her best suit. The examiners try to look suitably wise and severe. The relatives look suitably awed. The tone is quite combative until everybody says the Flemish (or whatever) equivalent of 'jolly good show' and a big party then goes on for hours. I was once asked to examine a thesis on Cambrian trilobites written by a young Sardinian whose relatives were all very small and well built. For all I knew they might have been brigands, and I sometimes wonder what would have happened if I had given the work a thumbs-down. As it turned out, it was the best party ever. In Britain, everything is low-key. The candidate sits in a small room with two examiners – one recruited internally from the university, the other an external examiner appointed by the examinations board. Blue jeans are perfectly acceptable. It is the external's job to give the candidate a hard time, and the grilling can, and often does go on for hours as the text is meticulously criticised. At the end, a cup of tea and a biscuit is in order. Minor corrections are usually deemed necessary before the thesis can be accepted, but it is rare to find a demand for a total revision. For my own examination I had an interesting time sparring with a young external examiner from Scotland, who became a friend thereafter. He was convinced by my arguments about the environmental controls on trilobites. But I did realise how necessary it would be to follow my professor's oft-repeated slogan about publication, lest somebody else had exactly the same ideas. I also appreciated that I should get those trilobites belonging to the family Olenidae in print as soon as I could because of their novelty – I could scarcely offer a scientific paper that referred to the crucial fossils as 'broad-brimmed Fred' and 'narrow-brimmed Fred'. I was admitted to my new degree at the earliest opportunity. Afterwards, Harry Whittington said 'Good morning, Doctor,' with a twinkle in his eye.

The fame of the Hinlopen Strait Ordovician fauna was spreading by that mysterious academic grapevine whose tendrils quickly grasp something of significance. The Norwegians must have felt that they had missed out on an important new discovery on their own Arctic bailiwick. The leading expert on trilobites of the family Olenidae was Professor Gunnar Henningsmoen of Oslo University. In 1957 he had published a splendid monograph on the Cambrian representatives of this group of fossils. It remains admirable today. What I had discovered was a hitherto unsuspected, much younger Ordovician evolutionary radiation of the same group of trilobites, a new chapter in the history of Henningsmoen's favourite animals. There were more than twenty species on Svalbard, none of them 'known to science' before. It would not be long before 'Fred' and his friends would be formally introduced to the world – it was extraordinary that they had not already been discovered elsewhere. In the rarified sphere of palae-ontology this was big news. I received an invitation to go to Oslo to present an account of the new discoveries at a meeting of the Norwegian Geological Society. I approached this public presentation with surprising sangfroid. I became much more nervous about such lectures in the following decades when I realised how they could influence my career, but at the outset I was buoyed up by a naïve bonhomie, an assumption that everyone before me in the audience was on my side. A goodly number of geologists awaited my talk in a rather austere wood-panelled lecture theatre. Several old men with whiskers could have been characters from a play by Ibsen. Projection slides clicked round on a carousel showing the bleak outcrops along Hinlopen Strait and a selection of my new trilobites. I explained why I thought this part of Svalbard was impor-tant for an understanding of the Ordovician world. When it came to questions a very tall, clean-shaven senior geolo-gist rose to his feet, and began in impeccable English: 'When I was in Novaya Zemlya in 1927 . . .' It was Olaf Holtedahl,* Norway's most famous geologist. I could

---

* Holtedahl must have been eighty-five at the time. He was a greatly respected figure, who was awarded the highest honour of the Geological Society of London, the Wollaston Medal.

scarcely have been more surprised had Charles Darwin himself appeared to ask my opinion. Another aged figure asked me what had happened to Vallance, and somewhat shamefacedly I had to admit I did not know. After the talk I was introduced to some important Norwegians who ran the Norsk Polarinstitutt, the body responsible for research in Arctic territories within the country's sphere of influence – which included the Svalbard Archipelago. They thought of themselves as the heirs to the great Roald Amundsen, and they did have command of plenty of resources for exploration. I met a young Englishman, David Bruton, another good trilobite researcher, who had settled a few years earlier in Oslo with his Norwegian wife Anne, and worked alongside Gunnar Henningsmoen at the Palaeontological Museum. It seemed that a large fraction of the local geological community had turned up for my talk. Something was afoot.

The following day I met up with Gunnar Henningsmoen and David Bruton at the old museum, which was reached by way of a pleasant walk to the middle of the University Botanical Garden. As in many museums, the research rooms were tucked away in the attic far from the public exhibitions. Gunnar proved to be as kindly as Harry Whittington, although he did smoke cigarettes with even more dedication than Professor Bulman. Indeed, everybody seemed to smoke at that time, including me; David Bruton, the archetypal tall, handsome Englishman, sported a perky pipe like that of a Second World War pilot. Through clouds of tobacco smoke I learned that my talk was part of the softening-up process to get sponsorship for a Norwegian expedition to Hinlopen Strait. The Polarinstitutt could handle all the logistics. A proper share of the trilobite booty would thereby be returned to Norway. This would not be an expedition on the lines of those organised for so many years by W. B. Harland of Cambridge University: no more dried meat bars and lemonade crystals. It would be a grander affair altogether, with proper field support. I realised at once that new collections would solve most of the tantalising questions that were still unresolved in my PhD thesis. After all, Geoff Vallance and I had been obliged to hurry through the entire rock sequence in a single season. There were rare and interesting species for which I lacked important information: I had heads with no tails and tails with no heads. I might find whole, beautiful trilobites of species I had only known from fragments. It would mean a delay in publishing the results, as new collections would require preparation, but there was no question that fresh material would help to make the eventual publications more authoritative. I had learned to photograph trilobites competently, but there was always room for improvement. And I had a job, even if I did not yet have tenure, so the pressure on getting into print was no longer quite so acute. No doubt a major expedition to Spitsbergen was a great opportunity to work with a different group of scientists. Three years had blunted the memory of the relentless, cold winds scouring the bare pebbles of the raised beaches along Hinlopen Strait. The cries of Arctic terns seemed romantic from this distance rather than heart-rending. The hours of freezing fingers clutching at sharp rocks as the geological hammer bashed on relentlessly were obliterated by the memory of an astonishing find or two made far to the north of the Arctic Circle. I could not wait to go back.

10

The Life Scientific

O n the shore of Hinlopen Strait in northern Spitsbergen a structure made from a few planks roughly hammered together sticks up from the endless plain. It has now partly collapsed, but enough remains to show it was once a cubicle-like affair built to give a little seclusion to the occupant, and to shelter him from the wind and the driving rain. A rectangular box in the centre giving into an exca-vated pit makes clear its original purpose: it was a privy for the least private place in the world. The whole thing is anchored on to a driftwood log dug into the featureless stretch of gravel to stop it blowing away in a blizzard. The *Playboy* pin-up that once adorned its walls has long since been shredded by the wind. It is the last of the objects in this book. In some sense it, too, is a fossil. Those few planks are almost all that remains *in situ* of the 1971 Palaeontological Museum and Norsk Polarinstitutt expedition to the Ordovician rocks of Ny-Friesland, Spitsbergen. The privy was well made: the photograph was taken when another

What remains of the 'room with a view' in Svalbard.

expedition visited the old campsite thirty years later. My own memories have been like that relic: the pieces are all there, but they have become disarranged through time and weathering, their shape and purpose preserved, but with many details dispersed or eroded.

The expedition had to coincide with the summer months, when perpetual light rules at 80 degrees north. Little had changed since 1967 – we still had to reach Svalbard by sea. The party left Oslo by train at ten in the morning on 14 July and arrived at Andalsnes at 16.45, and thence by bus to the port of Alesund, where we boarded ship at once.* After travelling north along the endless mountains that make

* The precision of these times might seem to belie my claims about my memory, but they are entirely thanks to David Bruton's disciplined diary entries, for which I am truly grateful.

the Norwegian coast we had to head towards the open sea, men stowed alongside all the expedition supplies. Once we left the safety of the fjords we had days to cross the fearsome seas northwards, an ordeal that tested even the best of sailors. Many members of the expedition turned ashenfaced and retired to their bunks for the rest of the trip. The Lofoten Islands were passed by in a blinding rainstorm. Further on, Bear Island loomed out of the mist, like an apparition. Our vessel was a decommissioned whaler taken on by the Norsk Polarinstitutt, the *Polar Star*. Fuel, food and canvas housing for the expedition was all loaded on to the tough old ship at Bodø and tied down with strong ropes to stop the crates shifting in high seas. When frightening waves towered over the bulwarks they would wash clean across amidships without threatening the buoyancy of the ship. We were then confined to quarters until the danger of being swept overboard had passed. For someone with good 'sea legs' like mine the principal problem was suddenly getting hurled against some unforgiving bit of steel as the vessel lurched unpredictably or crashed downwards having breached a precipitous wave. Eating was almost comical. Although tables in the mess were edged with low palisades to stop plates tipping wholesale into laps that did not prevent the plates from sliding about as the diner fruitlessly tried to catch up with his whale-meat stew. Potatoes – the Norwegian national dish whatever the claims of *fårikål* or *lutefisk* – could at least be speared if the diner were quick off the mark. Gravy always finished up washing about like slops. As the marine adventure continued towards the maelstrom the numbers dining fell progressively.

The working language was Norwegian, an important difference from the Harland expedition. I was a foreigner this time. Alongside his native English David Bruton had fluent Norwegian, and Gunnar Henningsmoen was perfectly bilingual, but the everyday chatter was naturally in the local tongue. I should say 'tongues' – despite its small population, Norway has many regional and local variants reflecting the isolation of the fjords in the past. Of all the Nordic languages, Norwegian is the most pleasant on the ear for a foreigner, musical and alliterative, at least as spoken by an Oslo chatelaine. It sounds much rougher from the mouth of a seaman hailing from Bergen or Tromsø. I was told that someone from Oslo could tell where any ordinary worker came from just by listening to his dialect. However, Norwegian is remarkably impoverished in swearwords. In fact, there seems to be only *one*: *'faen'*. It means 'the Devil' which sounds rather innocuous. In 1971, it was far worse than saying that other, old English four-letter word beginning with F, at least in polite society. If you hit your thumb with a geological hammer you might – possibly – be allowed to say *'faen'.* If you lost your fingernail in the process you might be allowed to say it twice. If you lost the finger – three times. The devilish Farne Islands off north-east England were doubtless dubbed by a passing Viking. Some of our deckhands were 'faening' away for all they were worth much of the time, a habit that was frowned upon by my more refined companions. Apart from the f-word I did pick up a little Norwegian as the expedition continued. I learned that many words had second cousins in English, often an old-fashioned form. It did not take a genius to recognise that a *'frisk bris'* was a fresh wind, as it was also a frisky breeze. *Brød* was obviously bread, and *knekkebrød* was easily recognised as the ubiquitous crispbread that has to be paired with sliced goat's cheese at breakfast. Other everyday words seemed to have no obvious relations – *'ost'* for cheese, *'smør'* for butter – unless the latter was somehow related to that buttery English word 'smear'. I soon learned that *'pass smøret'* was 'pass the butter' but I never did get to discuss the finer points of Ibsen's dramaturgy. For the most part, when I did not understand what was going on I cultivated an amiable silence. As always happens on expeditions, tension between incompatible personalities exploded periodically into arguments, but I kept out of trouble by smiling non-committally. I was regarded as a quiet person, which does not conform to my usual description.

Most nations have something to feel guilty about – a particular sin or weakness. In China, for example, it is gambling; when I first visited the People's Republic I recall the terrible punishments meted out to those who set up illegal gambling dens – and the money still squandered in Macao is legendary. In Britain it was, until quite recently, sex; the inhibited Englishman was a stock character. A whole genre of English theatre is built upon the embarrassment of trousers falling down. In Norway, it is alcohol. Anything with alcohol in it is inordinately expensive on the mainland. The argument advanced to me was that the population is at root Viking, and the people are likely to go berserk if allowed access to anything fermented. Sweyn Forkbeard lurks under the skin of every accountant in

Trondheim. Only by making beer and wine (let alone spirits) pricey is the nation kept on the straight and narrow – as opposed to the wobbly and wide. The route to Svalbard took us beyond the excise limits: booze was practically free! I began to see what my informant meant. There were drunks aboard ship. The worst was the cook, one Olav Stavard, who staggered around all day in a state of advanced intoxication. A couple of his friends were little better. He had to go, and go he did. I believe he was put off at Bodø in the far north of Norway, where we loaded supplies before our departure to the archipelago. The helicopter pilots could not be replaced, but they, too, were tremendous carousers. They even looked like Vikings. They did have one of the most dangerous occupations in the Arctic, so hard drinking was probably built into their survival strategy. The worst thing about them was that they insisted on singing. A few years after 'All You Need is Love' the hit of the day was a ditty beginning with the timeless lines 'I beg your pardon / I never promised you a rose garden'. This was the one tune they knew. The two pilots seemed to remember only the first lines, and these were repeated loudly and without end in a lurching way that somehow matched the motion of our vessel. Just when I might have fallen asleep 'I beg your pardon' started slurringly all over again. Since then, I have been careful to avoid this song, but it occasionally appears in bad dreams. Fortunately, the captain of the *Polar Star* was sober, a well-groomed figure with sleeked-back hair, and fond of sporting dark glasses. He did, however, wear carpet slippers on the bridge.

When the sea became calmer as we approached Svalbard

there was a chance to socialise in the saloon. I got to know some of those who would share the weeks to come. Gunnar Henningsmoen, a slip of a man and slightly bent, was serious in repose but had one of those smiles that instantly lit up his face. I discovered that he was epileptic; he was probably here against his doctor's advice. Relentless smoking probably did not help. He came with Frank Nikolaisen, his protégé, who was able to step in if Gunnar had a grand mal attack. Frank was short and stocky, with a strangely wide and narrow mouth that reminded me of E. H. Shepard's drawings of Mr Toad. He was from a humble background, but had taught himself to be an excellent trilobite researcher. I have never seen specimens dug out of hard rock with more precision than Frank could achieve. He was something of a man of mystery: he had business 'out East' in some part of the exotic Orient where he disappeared for long stretches of time, only to reappear and pick up his dissecting needle as if nothing had happened. He never explained what he had been doing or where he had been. It may (or may not) have been something to do with precious stones. Gunnar tolerated his strange lifestyle. On the sea crossing I joined Frank and two others for a game of bridge, and lost badly. Frank seemed to know all my cards. Then I was told he had once played bridge for Norway. Gunnar must have recognised his unusual gifts, and treated him rather like the wayward son he had never had. My Anglo-Norse friend David Bruton had several verbal spats with Frank during our field trip. They were temperamental opposites. I don't know what was said, but it caused an awkward silence in the mess tent and I knew it was a good time for me to look my most amiably bemused. A lichen expert, Haavard Østhagen, came along with us to sample the toughest organisms on earth; I remember his loud guffaws, so he must have been a humorous fellow. From the Norsk Polarinstitutt two experienced Arctic hands were aboard: Thore Winsnes and Thor Siggerud. Winsnes had been a pioneer geologist in the days of sleds and huskies; he was so skilled with ships that he took over from our cap'n on one occasion. He radiated calm. I had encountered Siggerud with Geoff Vallance on my earlier expedition. When I met up with him again he said: 'They say the bad penny always turns up.' I think it was a joke, but the fact that it has stayed with me for so long makes me wonder whether it was a value judgement. Was I *really* a 'bad penny'? I thought by now I was rather a good penny. I remembered those dismissive comments on my history of science essay from Nicholas Jardine in my Cambridge days. Praise seemed to wash over me, but every critical remark lodged forever in my heart. Why should I believe the latter rather than the former? Discuss, giving reasons, as my examination papers used to say.

Longyearbyen is Svalbard's largest town, and now has an airport and several hotels catering for those in search of the 'Arctic experience'. It has become a tourist destination, tucked down from the worst of the weather in the shelter of snow-clad mountain ranges. In 1971 it was still a frontier town divided by a single, rutted main road. It had grown up around coal mining, and the painted wooden buildings had a rough and ready appearance as if they were not prepared for the long term. In the summer thaw the town was muddy and unkempt. Cages holding huskies were scattered on the hillside, and the last of the fur trappers still turned up with their wild beards, and bundles of Arctic fox furs for the market. Longyearbyen was built for the export trade and ships of some size could safely dock there. It lies about halfway up the western side of Spitsbergen, sheltered from the wild ocean within a large fjord (Isfjorden); it was a relief to stand on dry land, even if terra firma continued to lurch in sympathy with the ocean for an hour or so until the body readapted. We then sailed on to Ny-Ålesund, which is billed as the world's most northerly town: then, it was just a few undistinguished wooden buildings, and a research institute and satellite tracking station at the top of a hill. Today it continues to be a centre for research on the Arctic and, of course, climate change. I remembered the terns reiterating their shrill, scolding cries, black guillemots and little auks bobbing on the waves, and eider ducks brooding motionless upon their camouflaged nests: little had changed since my first visit. As *Polar Star* steamed northwards from Ny-Ålesund we passed snow-capped mountains that made precipitous cliffs at the edge of the island, and we glimpsed where the ice sheets that covered much of its interior crept towards the sea. When an iceberg was calved from the front of a glacier a sharp crack like exploding dynamite was followed by a great splash, as the mass of ice settled into the water; as if it were alive, rather than simply obeying the laws of physics. *Polar Star* nudged most small icebergs aside. Passing around the north of the island we crossed the 80 degrees north line of latitude, an entirely unremarkable achievement for which tourists pay top dollar. The northern peninsula known as Ny-Friesland was generally less mountainous than the south of the island – bleaker, fringed with raised beaches of hard cobbles, and not a destination for the Zodiac dinghies that brought paying visitors to see walrus or Ross's gull. *Polar Star* progressed along Hinlopen Strait separating Ny-Friesland from the island of Nordaustlandet (Northeastland) to the east. We were approaching our outcrops of Ordovician rocks.

Supplies were ferried from *Polar Star* to the unforgiving shore: boxes of food and drink, tents, radio equipment, fuel drums, geological equipment, notebooks, everything we could possibly need. Rifles were on hand in case we were attacked by the *isbjørn*, the polar bear, which was capable of felling a human with one mighty blow. At supper-time there was always much joshing (in Norwegian) about *isbjørn*; if somebody complained of constipation a sudden encounter with an *isbjørn* was recommended as an instant cure. It was sad that we never saw a single *isbjørn*. The site may have been too bleak even for a polar bear. The camp was placed near the same melt stream we had used previously for our water supply. A large, communal tent was securely erected as dining room, office and kitchen. Sleeping quarters were shared tents, not much roomier than the one Vallance and I had lived in. David Bruton became my tent mate. He put up a Union Jack on a pole outside the tent; the Norwegian flag fluttered elsewhere, but it was better made than ours and stayed the course – the relentless wind blew ours to bits. After some unpleasant experiences digging our own toilet on the open beach the wooden latrine was constructed that still survives. The side facing the sea was left open so that the occupant could enjoy the view from the throne. Northern Svalbard is way beyond anywhere that trees grow but there was a quantity of wood lying on the beach, drifted in from Siberia, or from even further afield on the back of the North Atlantic current. Firewood was not a problem. *Polar Star* left us to our work, taking Winsnes and Siggerud on to some geodetic projects elsewhere in the archipelago. The Russians were after some of Norway's oil, and establishing the extent of territorial waters was top of the political agenda. When the sea ice blew in to Hinlopen Strait we were truly alone.

The routine of scientific collecting began. The weather was usually rather horrible, and blizzards occasionally prevented us from working at all. Northern Svalbard is the only place on earth where fog coexists with driving sleet. In perpetual daylight it is important to maintain a daily routine or the body clock goes haywire. Supper was on the table at the same time every 'evening'. There was no 'lights out' in the tent, so you hoped exhaustion would do the trick, and it usually did. I still have my eiderdown-filled Arctic sleeping bag – I cannot bring myself to throw it out. My feet were usually icy cold when I climbed into my snug bag, but the special feathers had a magical ability to nurture every particle of body heat, until, little by little, the toes began to thaw and then to gently glow. It was one of the best moments, and usually heralded sleep. Collecting filled most days, often just breaking hard limestone with geological hammers, putting promising material to one side for wrapping, while I made notes about the exact locality. Somebody pointed out that they used to make criminals break tough rocks for a punishment. We had aerial photographs, but no topographic maps. We were obliged to name features that had never been named before, and these labels had to be officially approved by some Norwegian government office. Vallance's and my 'Melt stream A' became Profilbekken (Profile Stream) for ever. It was not just trilobites that were immortalised. Several of the technicians from the Oslo Palaeontological Museum proved to be great assets in the field, working hard without complaint. Åge Jensen and Leif Koch hammered the rocks as if their own future depended on it. One or two of the others were less assiduous, blowing on their fingers against the cold, or overreacting if a chip of rock nearly went into their eye. When an exceptional trilobite turned up everyone gathered round to admire it. A specimen of the giant of the fauna – *Gog catillus* – split out perfectly the size of a small plate. For a while the collecting effort redoubled, but with hundreds of metres of rock to sample it is not surprising that not everyone was as fanatical as I was. Some parts of the rock succession were covered in ice that showed a reluctance to melt, and we dynamited it away to complete our sampling. Nothing was going to get in the way of telling the story of the rocks.

Although much of what the Oslo team did was following in the footsteps of the 1967 discoveries there was an area of exposed rocks that the earlier trip had never explored.

This outcrop lay to the south of the Valhallfonna glacier where it reached the sea in a massive cliff of pale blue ice. On the aerial photograph the dark rocks were quite obvious, outlined against the white ice sheet. To reach the exposure we had to cross in front of the ice face in our small, clinker-built dory. We felt like insignificant specks in comparison with the ice cliffs, from which small blocks regularly tumbled. We equally had to steer clear of ice floes, because the greater part of them was always treacherously concealed under water. Crazily eroded icy crags were scattered as if Henry Moore had had a sculpture retrospective floating at the top of the world. The new outcrop proved to be almost entirely developed in the black, deep-water limestone that included the olenid trilobites (*Balnibarbi* and *Cloacaspis* – 'Fred' – and many others) along with numerous graptolites. Nobody had ever collected there before, and the rocks were displayed in such a way that we could crawl over the surfaces of a series of sea floors that were 475 million years old. This was a fishing trip into ancient history, and we were the first ever fishermen. Even in the Arctic sleet it was enough to kindle a fire within a scientist's breast. It was clear we should spend more time recording and collecting from this virgin territory, which came to be called Olenidsletta (it donated its name to part of the rock succession). A small breakaway camp was set up on the shore so that we did not have to make repeated trips by boat. David Bruton and I, together with our field assistants Koch and Gram, used it as a base to make forays over the new territory.

One day I discovered a mistake in my PhD thesis. At the

end of my submitted work I had put together a trilobite from apparently matching, but separate heads and tails that were 'left over' after every other species had been assigned. I reasoned that they must belong together to make a completely new kind of trilobite. On a particularly frigid day on Olenidsletta I found a whole trilobite that proved I was wrong. In front of me was a superb, complete extinct animal addressing me from the rock surface in the very place it had laid down and died. It would be the first trilobite I named in the scientific literature – *Opipeuterella*. It had the mystery tail, but a completely different kind of head, so I had clearly failed to make head or tail of it before! I cackled into the icy wind when I made that bad joke to myself. I struggled back to the tent through the squalls to report my good fortune. Indeed, this little trilobite was previously completely unknown. It helped me realise that there were several trilobites with inflated, goggle-eyes that were probably part of a planktonic community rather than dwelling on the sea floor. I would spend much time over the following years telling their story. In a small way it was a eureka moment – and neither of my examiners had cottoned on to my mistake. I also learned an important lesson: it was OK to get things wrong. No piece of work is perfect. I was lucky enough to rectify my own mistake this time, but if somebody else had found out my error science would still have advanced, albeit in a modest way. Science is a process which moves towards a greater and more embracing truth, shedding its own mistakes, and this process binds together an internationally famous particle physicist with a young palaeontologist struggling to keep his cigarettes dry while whacking rocks in a blizzard.*

There were two skirmishes with death on the 1971 Spitsbergen expedition. I was lucky to survive both of them. The return to base camp from our outpost on Olenidsletta proved more hazardous than expected. My own recollections of the events are blurry, just a series of fleeting sensations. David Bruton was made of stronger stuff and, as on almost every day of his life, recorded what happened in his diary after the day's adventures. His entry for 9 August follows verbatim:

Two boats had planned to fetch us but one broke down and only one, skippered by Jensen, made it despite numerous ice floes in and around the beach. We soon found that we could not take both equipment and fossils so we left the latter to collect on a better day. We were Jensen, Nesland, Gram, Koch and Richard who sat in the stern while I sat high up on the cases with oars at the ready. The trim of the boat was pretty unstable and I said a silent prayer for a safe trip. The fog had come down again but we followed the shore and into the meltwater area in front of the glacier. The boat was going well in the following wind but each turn to avoid an ice floe, caused us to breach and the boat went precariously off balance.

Koch used the walkie-talkie to base and told them we were underway when suddenly a high-pitched rev of the

---

* That is not to say, of course, that the best scientist is the one who makes the most errors: that would be ridiculous!

outboard made me look astern to see the entire motor kicking violently in Jensen's hands. The transom board had come away completely from the gunwale leaving the motor to hang on the end of a short piece of safety rope tied to the stern ring. The motor stopped but not before I had both oars in the rowlocks trying to keep the boat from broaching to. It was almost impossible and I needed Koch's help with the starboard oar. We were drifting dangerously to an iceberg which rocked ahead of us and lashed by waves on the port bow. The boat was drifting right onto it and if I only could backwater the port oar and turn like mad with the starboard. I did, again with Koch's help and we shot by within touching distance of the blue ice front. Jensen, meanwhile, had succeeded in knocking the top plank of the transom into place again returning the nails into their well-worn holes. Would it hold? It did and soon the motor was going again and we shot past another ice floe which had loomed out of the fog dead ahead. Fortunately we were in smoother water and we could just see base camp. Even so, distances here are most misleading and it was an age before we were within what I judged to be swimming distance of the shore.

What David's succinct account did not mention was that men overboard are supposed to last for only about four minutes in Arctic waters before shock takes control. As for me, I sat in the stern doing nothing much to help the situation; possibly, I did not even realise that our lives were in jeopardy. I hope I was not just incompetent. Åge Jensen, the able technician from the Oslo museum, kept his head clear to avert a disaster that could otherwise easily have proved fatal. Only when we got back safely to base camp did David shakily confess how worried he had been.

My second escape from death did indeed involve immersion into a sea bobbing with ice floes. I was collecting from strata that dipped down into a choppy sea. Waves lapped up on to rocks that had been polished smooth by many years of marine scrubbing. Small ice floes jostled against one another further out to sea. I stepped down on to a sloping greenish surface – covered with tiny algae, I suppose – and my rubber boots had nothing to grip. I slithered uncontrollably into the deep icy waters beyond. This was it: four minutes. My trilobites flashed before my eyes. I recall with clarity the shock of the cold sea closing around me. I was wearing a good deal of woollen clothing and a heavy jacket; these would serve to drag me down. I gulped salt water in my confusion. It was my good fortune that David was working nearby. He quickly came to my rescue. Firmly braced so that he would not make the same mistake as I had, he reached out with his geological hammer. With one desperate lunge I managed to grab it, and somehow successfully scrabbled out of the sea, with David pulling on the handle end of the hammer. I was prodigiously chilled. The only thing to do was to run all the way back to camp to get changed into dry clothes. We were working a couple of kilometres away. So off I soggily set, relentlessly jogging through slushy ice and across melt streams and over stretches of loose cobbles. By the time I arrived at the communal tent the water in my wellington boots was lukewarm, through sheer bodily effort. Alone in the tent, the radio officer was making his usual unsuccessful attempts to contact *Polar Star*, while the radio whistled and crackled back at him like a lost alien. I am not sure he noticed my escape from death at all.

David Bruton became a good friend, and not just because I owe him my life. You cannot hide much if you are sharing a small tent during a twenty-four-hour snow-storm. We differed then in several ways. He was seven years older than I was, and had left England by the time the 1960s were beginning to offer a new version of society. Politically, I was far to the left at the time, and David was more on the right. Despite being the victim of rather typical Trotskyite manoeuvres while I was at King's College, my disillusion with communism was complete only when I visited Czechoslovakia a year or two later. David had been in Prague when the Soviet tanks rolled up in 1968 and had tales of an exciting escape from that intransigent regime; the International Geological Congress was being held there at the time. My old professor Harry Whittington had been at the Congress with David and must have been reminded of his earlier escape from the Japanese armies during the war. Both knew the impact of the iron fist, and they would not forget its heavy blows. I listened to David's account as we talked in our sleeping bags waiting for our feet to thaw. I guessed that politics were to be avoided if we were to live harmoniously, but I also had glimmer-ings of a greater tolerance of other people's views and backgrounds. We got along just fine. When the fieldwork was complete, the fossils were stowed, and *Polar Star* took us laboriously back to Norway. The helicopter pilots entertained us yet again with 'I beg your pardon'. For some reason it seemed more of an ordeal in this direction. When we reached Bodø on 7 September David gratefully left the ship as soon as he could, and continued his journey southwards by train. The collections found their way to the museum in Oslo much more slowly, where they were unpacked and registered. Most of them subsequently continued further to me in London for detailed preparation and proper study. I spent the next ten years of my life publishing the results and exploring their implications for science. Surely, this was the logical conclusion to that long journey from boy naturalist to scientific professional. What could go wrong?

*   *   *

When I arrived back in England I discovered that my wife had left me. While I was away in Svalbard she had moved back to the family home in Cambridge with Dominic, whose operations had been successfully completed. Back on familiar ground, she had developed a liaison with the husband of her old friend, an affair that had quickly developed into an overwhelming passion. By the time I arrived with my suitcases of dirty field kit I was almost irrelevant. I was the stranger who turned up on the doorstep, not welcome. As with the other traumatic events in my life, the details have been censored by some kindly overlord of my unconscious mind. All I have with me now is the feeling of being bereft. It was not supposed to be like this. This was not what the senior tutor had in mind for his research students. I was bewildered and hurt. I wanted to be with my son. It wasn't my fault; or else it was *all* my fault. I felt that I had not escaped the Arctic seas after all: I was swirling around in a dark ocean trying to stay afloat, way out of my depth. I felt anger, I felt sorry for myself. I felt both simultaneously, the way that squalls and fog could coexist on Hinlopen Strait. I can no longer recall where I went during the crisis. I believe I avoided going home; I did not wish to hear my mother's angry assertions that she had been right all along. I certainly imposed my misery upon my friends in Cambridge. I was a lost soul. All the small triumphs of discovery along Hinlopen Strait were eclipsed by heartache and confusion.

We tried getting the family back together, this time in a small apartment in Tufnell Park in North London. It was not going to work. I had lost my confidence, and my wife had moved on. The romantic days of my third undergraduate year seemed like an illusion, a trick of memory. The pretence that everything was back to normal soon fell to pieces. I packed up all my belongings into two suitcases, and became an itinerant, visiting on the weekends, a typical young, separated father who chased a small, laughing boy around the ponds on Hampstead Heath. I possessed almost nothing: a terrible old Austin Mini Countryman with a leaking petrol tank was my most valuable item. My wounds were invisible, but loss of trust is one scar that never heals completely. My senior sister-in-law kindly let me live in her unoccupied flat in Percy Street, in a raffish part of central

London north of Soho known as Fitzrovia. One side of the street was a rickety old Georgian terrace, and the flat was above an Italian restaurant and an out-of-hours drinking club run by a Mr and Mrs Con, who despite their name and occupation were eminently respectable. I could climb out through a skylight to hang out my smalls to dry between the chimney pots. This was my brief bohemian phase. I also discovered an inner-city ecosystem. The flat harboured cockroaches that lurked behind the sink and came out in some numbers at night. My mattress was on the floor, and I kept a shoe handy to whack the roaches if their scuttling woke me up. Once I awakened to find a mouse gorging on a fat squashed insect. This was a kind of balance between pests; for some reason I found it consoling. I suppose the message was: 'life carries on'. Some years later, after a relatively uncomplicated divorce, Bridget married yet another of my Cambridge friends and contemporaries, and Dominic went to live in France with his new family. Only then did the memory of my traumatic return from Svalbard begin to recede. A small but insistent voice told me that I had escaped from something that would have made me miserable.

My scientific work was set back. The Keeper of Palaeontology at the British Museum (Natural History) was compassionate enough, although my permanent appointment was put on hold, I assume to allow me time to prove my resilience. The new collections arrived from Oslo. Preparing trilobites from their rocky hiding place proved to be a kind of occupational therapy. I could not let my mind wander on to my wounds and worries because total concentration was obligatory – one slip and a valuable specimen was gone. A whole world was bound within the compass of a binocular microscope. I buckled down to months of work with 'vibrotool' and needles. I remembered Harry Whittington's advice: I had to get my new information about the olenid trilobites ('Fred' and allies) published before I could put out my theories about ancient continental boundaries. On the basis of the new collections, there were several species I had not known about when I completed my PhD thesis, and all of these needed work. Somehow, I wrote up the results of our new 1971 fieldwork, jointly with David Bruton, and published it in a mainstream journal of the Geological Society of America. My scientific publication list was growing, despite what had happened in my private life. There are many varieties of desperation, but only one kind of hope, and that is rooted in the future – in striving – in making a meal from a cockroach. Persistence would prove the stalwart virtue, more durable than idealism, and stronger than earthly passion: the imperative to carry on carrying on.

I began to expand my contacts. I soon discovered that one of the great things about science is that national borders mean little in a common endeavour. An advantage of being a specialist is that there are a relatively small number of colleagues around the world, and they soon become friends. Once my publications began to appear I sent reprints to people I had never met ('with compliments of the author' scribbled on the top) and received theirs in return. A pile of my reprints always arrived by post after any article was published, and these papers became the intellectual currency of my science. I started to build a library, in file boxes, alphabetically arranged by author. It signalled my entry into a community of like souls. Scientists working for the United States Geological Survey or at the American Museum of Natural History sent me copies of their works. We soon were on first-name terms. Famous Swedish specialists were generous with their celebrated contributions. Politics did not stop the flow of information. I received papers from Marina Chugaeva from the former Soviet Union. Doubtless, when I reciprocated, some KGB functionary in Moscow would have spent too much time trying to find secret messages concealed in trilobite language. Even China would emerge within a few years from the horrors of the Cultural Revolution, when it became obvious that some of the richest fossil faunas anywhere were to be discovered in the Far East. In a decade I would mentor a Chinese academic in London for two years. John Shergold and the great A. A. Opik became correspondents from Australia, where I would one day go to pursue my research. All my letters were typed on a machine that had already seen much service; staff were supposed to file a carbon copy of everything. When personal computers arrived I kept the old typewriter as a memento of more leisurely times. Letters could then be left for a week to mature; the peremptory responses demanded in the electronic era would have seemed fantastical.

Not everything was welcome: a proportion of published research proved to be inferior. A package from Monsieur Pillet in France was viewed with apprehension if it included smudgy photographs of poor specimens. Pillet published in obscure regional French journals, and there were doubts whether his papers had been through stringent peer review. Professor Techii Kobayashi was different. He was a senior Japanese professor and nobody *dared* challenge what he said in print. His illustrations were rarely adequate, and he coined new names with abandon. I was learning to pick my way through the minefields of scholarly reputation. London was, and remains, a hub for international travel, and some of my overseas colleagues would call in at the museum on their way to conferences or fieldwork, or to consult the collections. Then I could put a face and a handshake to the reprints. I began to feel that this world was where I belonged, but it was not the world of the Natural History Museum alone. I had joined a community of enthusiasts from almost every continent.

When my short time in Fitzrovia came to an end I lodged with another friend whom I had first met on the geological field trip to Arran in my first year at Cambridge University. I was drawing away from Bridget's family. Andrew Sita Lumsden's career took a biological turn, and eventually he ran a famous Medical Research Council laboratory attached to Guy's Hospital in London. He was elected to the Royal Society well before I was. He had a brief spell as a graduate vertebrate palaeontologist at Yale University, but that did not end happily, although he met his beautiful wife Ann while in the USA. In the early 1970s he looked like a slightly epicene rock star, with hair – a little lank – well below his shoulders and clothes that belied his handsome face and posh accent. He was a motorcycle fanatic, and also an accomplished engineer. The front room of the Victorian terrace in MacKenzie Road, Penge, housed a gigantic lathe on which metal thingumajigs could be turned with precision. Andrew's bike, a Vincent, was set up on a kind of altar, ready for worship. I once went for a ride on the pillion that was hardly less terrifying than falling into the Arctic Ocean. Andrew acted as a kind of consultant for the South London biking community, and tough-looking types in leathers were wont to turn up at the front door, waving broken metal whatsits for repair. Andrew and Ann welcomed the man with two suitcases into their spare room – 'damaged goods', my mother would have said. I wondered if any woman could ever be interested in someone who owned only responsibilities and a Mini Countryman with a leaking tank.*

* * *

In this book I have emphasised how twists of fate – luck if you prefer – repeatedly tweaked my story, allowing me to follow a unique route that led to an office deep within the Natural History Museum in London. My tale would have been different had a crucial toss of the die fallen any other way. If I had made another intuitive response to A. Sainsbury-Hicks my life would have been directed towards the humanities rather than science. Harry Whittington might well have stayed at Harvard University. I might have decided not to go on that first Spitsbergen adventure, and

---

* Fortunately, there was, but that is not a part of this book.

there may have been nothing special about the rocks when I got there, or Geoff Vallance might have had another fifteen marks in his final examinations. Bill Dean could have become head of palaeontology at the Natural History Museum – and there would have been no job. The list of my lucky chances is impressive, and has nothing much to do with any gift of character or intelligence on my part.

A sceptical reader might wonder with some justice whether luck alone may provide what philosophers call 'necessary and sufficient' conditions for prospering in science. I am sure luck plays a part, but knowing how to put good fortune to equally good use is just as important. The proverb that 'genius is an infinite capacity for taking pains' is all very well, but a mountain of observations is not a hypothesis. Imagination operates alongside hard work, and both can be seeded by a fortunate turn of fate. I had to make some intellectual leaps to link what I had discovered in Svalbard to advances in understanding in plate tectonics. Again, I was lucky enough to be in the right place – Cambridge – at the right time, but I could have earned a perfectly respectable doctorate without engaging my imagination in any kind of gymnastics. Luck, I believe, is a boon that must not be squandered. Luck gifts foundations but does not build a castle. If a theory can be proved by its converse, I know of examples where fortune has been generous but its magnanimity has been wasted. In the mid 1970s I spent a year in Newfoundland, off eastern Canada. I learned of some astonishing fossils that had been discovered at Mistaken Point, on the eastern side of the island. They were in what were then called Precambrian strata, and at that time the discovery of an array of large, frond-like fossils this ancient (*c*.565 million years old) was sensational. The original finds were made by a geology student called C. B. Misra in 1967, but the site was regarded as the 'property' of a local academic at Memorial University, Mike Anderson. From time to time he would show casts of the extraordinary fossils, always promising to reveal them fully to the world. He never did. He failed to do justice to an extraordinary gift, but kept other researchers away. After Anderson's retirement, the Mistaken Point fauna was proved through publication to be seminal to our understanding of early life, and its locality is now a UNESCO World Heritage Site.

Science is one of the few areas of human endeavour where merit is still the main criterion for success. 'It's who you know, not what you know' does not apply. Science does not listen to accents; nor does science care about skin colour. Women are still outnumbered by men, but there is nothing intrinsically sexist about science, and the numbers are changing as ways of balancing work and family are developing. Very few scientists become rich through science, but then most of them are not in it just for the money. Science is not a moral system – it is a method of discovery, and has no direct part to play in organising society, although society is obliged to listen to scientific evidence to make up its mind. Too few scientists engage with politics; they tend to be suspicious of murky compromises and wary of how scientific evidence is ignored in favour of short-term gains. Those scientists that *do* serve government tend to leave the laboratory far behind. Twenty-first-century politicians often perceive public science as little more than a way of feeding innovation into the economy – for every dollar spent, two dollars earned. It is as if the job of the scientist were regarded as churning out profitable patents. This is not good news for those studying mimicry in butterflies or, for that matter, those fascinated by Ordovician trilobites. Sometimes I consider my luckiest break of all was to be born at exactly the right time.

*   *   *

The International Geological Congress is held every four years, when all manner of scientists involved with the earth and its history gather for a series of sessions spread out over a week. Titles and abstracts are offered, and papers presented in front of leading specialists spanning the spectrum of geological subjects, palaeontology included. It is a big affair. The 1968 Prague Congress had been abruptly truncated by the Russian occupation of the city, when David Bruton and Harry Whittington had made good their escape. No such troubles threatened the 1972 Congress to be held during the summer in Montreal, the most populous city in Quebec. The British Museum (Natural History) had agreed to finance my first transatlantic trip. I was to represent the BM at the IGC! Even the initials gave the trip a kind of special imprimatur: after all, I was part of an exclusive club that actually knew what they meant. Better still, the meeting was preceded by a field trip across the Great Basin of Utah and Nevada in the western United States. The rocks there had yielded several trilobites of the same age and kind as the ones we had collected and identified from Svalbard. This was my first outing in a twenty-year mission to chase rocks of this Ordovician age around the world to test out theories about ancient continents and trilo-bites. So I enrolled with TIGBOE, the Third International Great Basin Ordovician Expedition: I was going to IGC via TIGBOE on behalf of the BM. It could have been a secret code.

Portrait of the artist as a young geologist.
With Lehi Hintze (left) in Nevada in 1972.

The expedition started from Denver, Colorado, and moved westwards over the Rockies. The organiser was Reuben James Ross Jr of the US Geological Survey, who had mapped these rocks and collected their fossils throughout his working life. His monograph of twenty years earlier was a classic. 'Rube' was, in the demotic of the region, a somewhat ornery character, but he was charm itself on TIGBOE, apart from rallying everyone at 5.30 in the morning for a useful start. The party was an international crew – from France, Russia, Norway, Sweden, Denmark and Australia. I came with Robin Cocks from the Natural History Museum, and struck up a friendship and collaboration that has lasted a lifetime. Our line of all-terrain vehicles moved from range to range across the west. We lodged in Nevada casinos and cheap motels. The spacious and beautiful landscape was divided into high ranges – still partly snow-clad in early summer – separated by great arid basins floored by endless, aromatic sagebrush. The flats of playa lakes occupied the lowest ground, where water gathered and evaporated to perfectly smooth salt surfaces that glistened white in the sunlight. The only trees were pinyon pines scattered high up on the ranges. The whole landscape was fragrant from top to bottom. There was nobody out there. All roads ran straight across the basins, and then twisted and turned to cross the ranges; many of them were unmetalled, and their gravel surfaces developed corrugations known as 'washboard' that mercilessly rattled field vehicles. Coffee was spilled into laps.

The Ordovician rocks were exposed on the flanks of the ranges and wonderful sections through the strata lay on every side. The only problem was what to sample first. For me, it was the validation of my hard-won expertise. I was first out of the vehicle, and last back at the end of each stop. I scurried up steep slopes in a frenzied attempt to collect as much as I could in the shortest possible time. The abundant trilobites were old friends from Spitsbergen, I knew many of them better than did the older and greyer experts from around the world. I could already see that many of the ideas I had developed on the Arctic wastes could be applied here in the semi-desert – during the Ordovician period they had both been part of the same ancient continent, flooded by warm, shallow seas full of life. We were joined in Utah by Lehi Hintze, another trilobite expert – and a Mormon. He wore long woollen underwear as demanded by the Church of the Latter Day Saints, which seemed an extraordinary punishment as the heat built up; neither was he allowed to cool down with an icy beer at the end of the day. We rubbed salt around the rim of the glass to help make up for what we had sweated out during the heat. Lehi swigged a Coke, and even that was considered naughty. I rehearsed some of my ideas to him and he smiled a very sweet smile; I am still unsure whether he believed a word I said. One evening, I had the chance to present the talk I was planning for Montreal to the whole party. After sunset, a sheet was rigged up as a substitute for a white screen, and a slide projector was somehow powered from the vehicle. There was nothing but empty, open countryside for miles around, and a million stars made a distant ceiling. In the wild American west it seemed curious, but somehow wonderful to show images of David Bruton and myself with a backdrop of icebergs. I explained my ideas about the different habitats occupied by trilobites in relation to past Ordovician continents. By now, I felt I was among friends. My arguments seemed spun out of pure logic. An owl hooted approval from somewhere far away. For the first time in months I felt buoyant.

Return to civilisation was sobering. It was hard to believe that there could be so many geologists in the world as were strutting around Montreal. The 24th International Geological Congress embraced ranks of geophysicists, bearded tectonics wallahs, oil men, stratigraphers, surveyors, glaciologists, petrologists ... and some palaeontologists hidden away somewhere in the melee. After a plenary session given by the grandees, the crowd would separate into their specialist disciplines for the rest of the meeting. But first, the 'icebreaker' was a vast congregation of chattering geologists; old friends were greeted with a vigorous pumping of hands, Russian delegates tried to evade their KGB shadows, people from Far Eastern countries tried their best with English, which did not do them much good when addressing a Québécois. It was hard to read the badges that hung around delegates' necks (some years later I was labelled 'Richard Farty' thanks to a careless computer). In Nevada and Utah I had felt like one of the gang, a member of a special club. At IGC I realised that I was a small fish in a very large pond, and almost nobody knew who I was. This was the world in which I had chosen to make my mark. My new French pal Jean-Louis Henry and I wandered around with Gauloises dangling from our lips trying to look like Jean-Paul Belmondo. Robin Cocks was a few years older than I was, and introduced us to some of the friends he had made from previous meetings. Rube Ross made sure we met his colleagues from the Geological Survey. The word 'networking' had not been invented, but that was what I was doing. In the end, I thought the throng might be negotiable. Robin and I sloped off to the room we shared in one of the more ordinary Montreal chain hotels. I felt nervousness creeping on as I tried to get to sleep. The air conditioning hummed off and on, doubtfully.

The small, modern lecture hall was quite full the following day. The palaeontology sessions had brought together my colleagues from around the world, while the other geologists attended their own lectures in adjacent venues. I could recognise a number of friends from the Great Basin expedition, but there were many more dele-gates that I had never seen before. Some were quite elderly and looked a tad fierce. I guessed that I was probably the youngest person in the room. Before computers made mechanical devices redundant a slide projector clunkily projected images on to a screen behind the speaker. Slides had to be inverted to appear right way up for the audience, and I had checked mine repeatedly before inserting them into the carousel. It would have been unthinkable to show our field tents dangling from the sky! I was preceded by several other talks: the chair introduced, the paper was presented, questions were fielded, the session moved on. Abstracts were already printed in the programme. I was in a daze, twitching internally. The Russian speaker before me had brought thick glass projection slides. The lamp in the projector was too hot and the glass cracked onscreen; the poor lady had to see her images covered with a spider's web of cracks that made them almost indecipherable. What if some disaster happened to me as well? The butterflies in my stomach rehearsed a major migration. It was my turn. The title was announced by the chair – and the slides were fine. I whizzed through my talk efficiently enough, but without the assured pizzazz I had displayed a few days earlier. There was brief applause. The person who always asks the difficult question asked the difficult question. Somebody made congratulatory noises. I thought I caught the word 'garbage' coming from one of the grizzled figures to my left: clearly, not everyone was impressed. I returned to my place in the auditorium, hugely relieved. The ritual moved on. Attention switched to the next topic. I had survived. I had personified my ideas, and now I could be just another delegate. The new boy from the BM had been introduced to the world. All the other boys I had once been listened to my words from very far away. I was, at last, a scientist.

Acknowledgements

Arabella Pike, my long-term editor at HarperCollins, suggested that I write this memoir of my early years. Without her generous invitation this book would have remained unwritten. As so often, Heather Godwin read and improved my first version; if a few feeble jokes fell to her perceptive axe it was in a just cause. My wife Jackie once again tolerated the selective amnesia of the writer at work. My gratitude to all three is beyond measure.

For details of my story I tested the memory of my sister, Kath, and now I can apologise for what she endured at the hands of her older brother. My old geological friend David Bruton kept diaries when I failed to do so, and generously allowed me to quote from his record of our adventures in Spitsbergen. My father's early years were elusive, but I discovered what an outstanding sportsman he was by contacting Worcester Royal Grammar School; the school secretary Joanna Weaver provided much evidence of his prowess as recorded in old numbers of the school magazine. This was most helpful. My father's fishing achievements were equally remarkable, and I must record my gratitude to Peter Hadwin of the Watford Piscators for looking into records from the River Gade. It seems my father's trout record may still stand. My old school, Ealing Grammar School for Boys, no longer exists, but the old Ealonians do, for which I am thankful. I reproduce herein a sketch of Forge Cottage drawn by one of my school contemporaries, and I wish I could be precise about its attribution. Two of my schoolmasters – John Railton and K. E. Williams – have earned my thanks for significantly shaping my life. Both of them are no longer alive, but that does not diminish the gratitude they deserve. Friends from my schooldays, Bob Bunker and Robert Gibbs, have recalled the curious world of the grammar school, and stimulated my own recollections. From my university days, Victor Gray and Michael Welland prompted memories of my time as a young dog, and I am saddened that Michael died before he was able to read my stories. Clive Wilmer helped revive my young poetical self.

Rob Francis provided the photographs for this book, and I am indebted to his skill as a photographer. Jackie Fortey and David Milner trawled the manuscript for the errors that passed unnoticed by the writer. I never kept diaries, so despite my best efforts there will be errors in my account of distant decades. I admit full responsibility for any such shortcomings.